S276 Geology
Science: Level 2

Book 2
Magmas and Mountains

Written by Stephen Blake (Book Chair), Tom Argles and Nigel Harris, based in part on earlier contributions by David Rothery and Andrew Bell.

This publication forms part of the Open University course S276. Details of this and other Open University courses can be obtained from the Student Registration and Enquiry Service, The Open University, PO Box 197, Milton Keynes MK7 6BJ, United Kingdom (tel. +44 (0)845 300 60 90; email general-enquiries@open.ac.uk).

Alternatively, you may visit the Open University website at www.open.ac.uk where you can learn more about the wide range of courses and packs offered at all levels by The Open University.

To purchase a selection of Open University course materials visit www.ouw.co.uk, or contact Open University Worldwide, Walton Hall, Milton Keynes MK7 6AA, United Kingdom for a brochure (tel. +44 (0)1908 858793; fax +44 (0)1908 858787; email ouw-customer-services@open.ac.uk).

The Open University
Walton Hall, Milton Keynes
MK7 6AA

First published 2010.

Edited and designed by The Open University.

Typeset by The Open University.

Printed and bound in the United Kingdom by Halstan Printing Group, Amersham.

The paper used in this publication is procured from forests independently certified to the level of Forest Stewardship Council (FSC) principles and criteria. Chain of custody certification allows the tracing of this paper back to specific forest-management units (see www.fsc.org).

ISBN 978 1 84873203 2

1.1

The paper used in this publication contains pulp sourced from forests independently certified to the Forest Stewardship Council (FSC) principles and criteria. Chain of custody certification allows the pulp from these forests to be tracked to the end use (see www.fsc-uk.org).

The S276 Course Team

Course Team Chair
Peter Sheldon

Course Manager
Glynda Easterbrook

Main Authors
Tom Argles
Stephen Blake (Book 2 Chair)
Angela Coe
Nigel Harris
Fiona Hyden
Simon Kelley
Peter Sheldon (Book 3 Chair)
Peter Webb (Book 1 Chair)

External Course Assessor
Dr Alan Boyle
(University of Liverpool)

Other Course Team Members
Kevin Church (Consultant Reader)
Roger Courthold (Graphic Artist)
Sue Cozens (Warehouse Production Manager, Home Kit)
Sarah Davies (eLearning Advisor)
Ruth Drage (Media Project Manager)
Linda Fowler (Exam Board Chair)
Michael Francis (Media Developer, Sound and Vision)
Sara Hack (Graphic Artist)
Chris Hough (Graphic Designer)
Richard Howes (Lead Media Assistant)
Martin Keeling (Picture Researcher)
Jane MacDougall (Consultant Reader)
Clive Mitchell (Consultant Reader)
Corinne Owen (Media Assistant)
Andrew Rix (Digital Kit video filming)
Colin Scrutton (Consultant Reader)
Bob Spicer (Reader)
Andy Sutton (Software Developer)
Andy Tindle (Digital Kit and Virtual Microscope photography)
Pamela Wardell (Editor)

Course Secretary
Ashea Tambe

Other contributors are acknowledged in specific book, video and multimedia credits.
The Course Team would also like to thank all authors and others who contributed to
the previous Level 2 *Geology* course S260, from which parts of S276 are derived.

The cover photograph shows Rackwick Bay on the island of Hoy, Orkney. The boulders are mostly Devonian sandstones and conglomerates of the Old Red Sandstone, interspersed with dark-grey, generally smaller, boulders and cobbles of basaltic lava, also of Devonian age. Copyright © Andy Sutton.

Contents

Chapter 1 Introduction

By this stage in the course, you should be able to make sense of some basic geological materials. We hope you can now look at a geological map or cross-section and interpret it in terms of the three-dimensional distribution of the rock units shown, and examine a hand specimen or thin section and make some deductions about its composition and origin. But don't worry if you are not yet feeling confident with these skills; there are more opportunities to practise them in this book. Whereas Book 1 dealt mainly with the 'building blocks' of geology (e.g. minerals, rocks, maps and cross-sections), Books 2 and 3 concentrate on geological *processes* – the processes that produce and re-arrange rock materials and shape the Earth's ever-changing landscape and internal structure. Using a combination of observation and reasoning, and a knowledge of geological processes that can be seen happening today, rocks can be interpreted in terms of the ancient environments and events they represent. In this way, we can deduce what it would have been like to have been there at the time – one of the main aims of geology.

Some geological processes can be readily seen going on around us, such as the erosion of shorelines and the construction of coral reefs. On the Earth's surface, rocks and minerals interact with water, air and life (more generally the hydrosphere, atmosphere and biosphere). Book 3 will look at these surface processes and interactions, which are recorded by sedimentary rocks and the fossils they contain. But the Earth's surface also reveals many signs of *internal* activity – particularly erupting volcanoes, huge areas of igneous and metamorphic rocks, and rocks that show evidence of distortion by powerful forces (Figure 1.1). These rocks owe their origins to processes happening within the Earth. It is these processes – the ones responsible for producing magmas and volcanoes, and building the Earth's highest mountain ranges – that are the

(a)

(b)

Figure 1.1 Magmas and mountains formed by the Earth's internal processes. (a) Lava on Hawaii in the process of flowing, cooling and solidifying into volcanic rock. Notice how the surface shapes reflect the behaviour of the red incandescent magma as it moves. (b) Deformed strata, exposed on a mountain face in the Swiss Alps, which have been changed from their original horizontal orientation and been folded back on themselves during a mountain-building event.

subject of this book. We will deal with the areas of geology concerning volcanic eruptions (volcanology), the characteristics and interpretation of igneous and metamorphic rocks (igneous and metamorphic petrology), and the study of deformed rocks (structural geology).

We begin with a brief review of the Earth's structure and an outline of plate tectonics (the idea that the Earth's outer surface is composed of enormous slabs of solid rock, each of which is slowly moving, causing geological activity, especially at plate boundaries). This will provide a context for looking in more detail at the Earth's internal geological processes and the rocks they produce.

Chapter 2   The solid Earth

2.1   Setting the scene

The Earth is the largest of four planets in the inner part of the Solar System. These planets are similar in terms of size, mass and density (and composition) and constitute the **terrestrial planets** (Table 2.1). They are much less massive, though denser, than the giant planets Jupiter, Saturn, Uranus and Neptune. All of the planets in the Solar System are about 4.6 billion (4600 million) years old.

Table 2.1 Basic data for the terrestrial planets. The Moon is not a planet in the astronomical sense, because it orbits the Earth rather than the Sun; however, it has the geological attributes of a terrestrial planet and is therefore included in the table. (For the same reason, one of Jupiter's satellites, Io, is sometimes classified as a terrestrial planet.)

Name	Average radius/km	Mass/10^{24} kg	Density/kg m^{-3}
Mercury	2440	0.3302	5427
Venus	6052	4.869	5243
Earth	6371	5.974	5515
Moon	1737	0.07349	3350
Mars	3390	0.6419	3933

Question 2.1

The densities of rock types collected near the Earth's surface range from about 2.3×10^3 kg m^{-3} (sandstone) to 3.3×10^3 kg m^{-3} (peridotite). The average density of the Earth quoted in Table 2.1 falls well outside this range. Give two explanations that could account for the discrepancy.

Actually, the Earth's average density is believed to result from a combination of both the effects described in the answer to Question 2.1. As depth (and hence pressure) increases, rock becomes compressed, but this effect causes only slight increases in density (except where particular pressures are reached that allow certain minerals to undergo phase transitions and change to denser structures). The main reason that the density of near-surface rocks is much less than the Earth's average density is that the Earth's central part (its core) is not made of silicates at all, but is made of iron-rich material that is more than 1×10^4 kg m^{-3} in density.

Based on this argument, all the terrestrial planets are believed to have iron-rich cores, with the possible exception of the Moon, but far more is known about the Earth's interior than that of any other planet because it has been mapped by studying how seismic waves (vibrations triggered by earthquakes or explosions) travel through it. There is a great deal of seismic and other geophysical evidence bearing on the Earth's interior, but for the purposes of this geology course only the most relevant results are reviewed in the next section.

2.2   The Earth's internal structure

Viewed on a global scale, the Earth's interior is composed of concentric layers. Some of the layers are defined on the basis of chemical composition, while other layers are defined by their mechanical properties.

2.2.1   Compositional layering

The most fundamental compositional distinction within the solid Earth is between the **core**, which occupies the centre and is iron-rich, and the overlying rocky material of the mantle and crust (Figure 2.1). Study of the way seismic waves propagate from one side of the Earth to the other shows that the core consists of inner and outer parts. The inner core is solid and has a composition dominated by iron with a little nickel. The outer core is liquid and consists of molten iron diluted by around 5–15% of less-dense elements, most likely one or more out of oxygen, sulfur or silicon. Motion within the fluid outer core (which is electrically conducting) is responsible for the Earth's magnetic field.

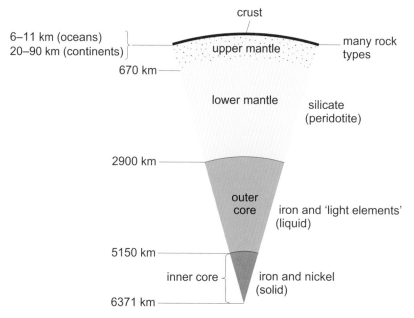

Figure 2.1   The compositional layers within the Earth, showing their depth and composition. At each boundary the seismic wave speeds change suddenly, indicating a change in composition or mineralogy. The thickness of the crust has been exaggerated for clarity.

Above the core lies the **mantle**, which makes up most of the remainder of the Earth. The distinction between core and mantle is very clear cut, because the mantle consists of silicate minerals and has a bulk chemical composition equivalent to that of the ultramafic rock type known as peridotite that you met in Book 1, Section 6.3. However, this is not to say that the whole mantle consists mostly of olivine and pyroxene.

■ What is the reason for this?

☐ It is because, at sufficiently high pressures, minerals become so compressed that they can undergo a phase transition to a denser atomic structure. They can turn into different minerals.

A notable phase transition in the mantle is believed to occur at about 410 km below the surface where olivine is compressed to a denser mineral with the structure of spinel (Book 1, Section 4.7.2). At still greater depths of about 670 km the spinel phase transforms to a mineral called perovskite, which has an even higher density structure. This phase transition is the likely explanation for the change in seismic wave speeds that marks the boundary between the upper and lower mantle (Figure 2.1). However, these phase transitions do not accompany changes in the overall chemical composition of the rock.

The outermost solid part of the Earth is the **crust**. The distinction between crust and mantle is less fundamental than between mantle and core. Crust and mantle are both predominantly composed of silicate minerals, but differ chemically in that the crust has a higher percentage of silica (SiO_2) and alumina (Al_2O_3) than the mantle (Table 2.2), and so is made of less-dense rock with lower seismic wave speeds.

■ What is the percentage of SiO_2 in an ultramafic rock type such as peridotite?

☐ Ultramafic rocks are defined as consisting of <45 wt % SiO_2 (Book 1, Table 6.4).

In fact, the mantle is thought to contain approximately 43–46 wt % SiO_2, whereas the average SiO_2 content of the crust is about 10% higher. The picture is complicated because the Earth has two distinct types of crust, of differing compositions. **Oceanic crust** (about 49 wt % SiO_2) forms the floor of the deep oceans. **Continental crust** (about 60 wt % SiO_2) makes up, as you might expect, the continental land masses and also the floors of the shallow **shelf seas** that cover the **continental shelves** around the edges of most continents.

Table 2.2 Estimated average composition of the mantle and the two types of crust, expressed as weight % oxide, and their approximate average densities at low pressure.

	Mantle	Oceanic crust	Continental crust
SiO_2	45.4	49.5	60.6
TiO_2	0.21	0.90	0.72
Al_2O_3	4.5	16.8	15.9
$FeO + Fe_2O_3$	8.1	8.1	6.7
MgO	37	9.7	4.7
CaO	3.7	12.5	6.4
Na_2O	0.35	2.2	3.1
K_2O	0.03	0.07	1.8
density/10^3 kg m^{-3}	3.3	2.9	2.8

Question 2.2

To which of the rock groups – ultramafic, mafic, intermediate and felsic – are the average compositions of (a) oceanic and (b) continental crust in Table 2.2 most similar?

It is important to remember that Table 2.2 shows average compositions only. The average composition of the oceanic crust is actually quite representative of the ocean floor, where it is rare to find extensive tracts of igneous rocks whose composition is other than mafic. However, continental crust is much more variable in composition and there are large volumes of continental crust, particularly in the upper part, where the composition is felsic rather than intermediate. The continental crust is geologically complex, with many different sedimentary, metamorphic and igneous rock types present, making it particularly difficult to estimate its global average composition, so you may find values quoted elsewhere that differ by several per cent from those here. However, there is no dispute that continental crust is broadly speaking intermediate in composition and thus significantly richer in SiO_2 than the mafic oceanic crust.

Where continental crust is joined to oceanic crust, there is a lateral transition from one type to the other (a **continental margin**) that may extend over tens of kilometres. In contrast, the junction between crust (whether oceanic or continental) and the underlying mantle can usually be located precisely because it is marked by a jump in the speed at which seismic waves are transmitted. This discontinuity in seismic speeds defines the crust–mantle boundary and is termed the **Mohorovičić discontinuity** (named after the Croatian geophysicist Andrija Mohorovičić (1857–1936), who discovered it in 1909), usually abbreviated to the **Moho**. The average thickness of the oceanic crust is about 7 km, whereas the average thickness of the continental crust is about 38 km. The latter average value hides significant variation and the thickness of the continental crust is greatest in areas with high surface elevation. For example, the Moho lies up to 80 km beneath the Bolivian Altiplano and up to 90 km below the Tibetan plateau and the Himalaya.

2.2.2 Mechanical layering

Think about the properties of the rocks at the Earth's surface. An obvious property is, perhaps, that they are brittle – rocks break when hit with a geologist's hammer. On a larger scale, rocks break to release an earthquake when they are subjected to large enough forces. Deeper in the Earth the rocks are hotter and, just as warming a candle or toffee changes it from brittle to malleable, they can become hot enough to flow slowly rather than break. It is therefore possible to think of the Earth as having a mechanically strong outer shell, consisting of the crust and the top few tens of kilometres of the mantle, which forms a discrete mechanical unit called the **lithosphere**. *Lithos* means 'rock' in Greek, and the term lithosphere was invented to describe the shell that behaves like rock in the familiar sense of being rigid. The lithosphere is typically about 100 km thick, but is thinner beneath parts of the ocean and can exceed 200 km beneath ancient continents.

Below the lithosphere lies the mechanically weak part of the mantle, known as the **asthenosphere** (*astheno* is Greek for 'weak'). The crystalline rock material forming the asthenosphere, although chemically and mineralogically similar to lithospheric mantle, behaves mechanically like a very slowly moving fluid because the atomic lattices within the crystals can distort at the prevailing pressure and temperature in response to applied pressures. The rock need not be hot enough to melt for it to behave in this way. There is one region of the mantle, at between about 50 and 200 km depth and variously known as the low-speed layer or **low-velocity zone**, where seismic wave speeds are a few per cent lower than normal, and this is probably caused by a small amount of melting. Movement between crystals lubricated by this liquid will also allow the mantle to deform by flowing rather than by breaking.

Different Earth scientists use the term asthenosphere in different ways. Some use it to refer to all the mantle between the lithosphere and the core, whereas to others the asthenosphere is seen as the same thing as the low-velocity zone. A middle option confines it to that part of the mantle from the top of the low-velocity zone down to the 670 km interface between the upper and lower mantle. A precise definition is unnecessary here, because this course is not concerned with the behaviour of the deeper mantle. The mechanical layering of the crust and upper mantle is summarised in Figure 2.2.

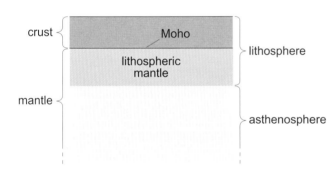

Figure 2.2 The division of the crust and upper mantle into two mechanical layers: the lithosphere and the asthenosphere.

There are several processes that cause the mantle lying beneath the lithosphere to flow. For example, heat emanating from the core warms the lowermost mantle, making it slightly less dense. As a result, the warmed mantle rises and is replaced by cooler, denser, mantle. This is the process of convection and involves the slow flow of mantle material in response to density differences caused by heating.

Geological processes at the Earth's surface can also generate forces that cause the mantle to flow. When a large volcano builds up on the surface, when a high mountain belt is formed, or when a thick sedimentary deposit or ice sheet accumulates, the mass of material pressing down on the mantle increases. In the manner of a ship that is being loaded, the extra weight forces the lithosphere down into the fluid asthenosphere. Likewise, if a thick ice sheet melts, the lithosphere will rise (just as a boat rises higher in the water when it is unloaded). The result of this effect can be seen in northern regions of Canada and Europe that were buried by up to 3 km of ice during the last major glaciation about

80 000 to 10 000 years ago. When the ice melted, the land started to rise; shorelines receded out to sea so that ancient beaches were left high and dry (Figure 2.3a)[1]. Although the ice sheets have long since gone, there are places where coastal moorings have had to be abandoned as the shore still continues to recede. Measurements of mean sea level on the Baltic coast were first made by the Swedish scientist Anders Celsius (1701–1744) (of temperature scale fame) in the 18th century and have since revealed that the land is still rising by about 1 cm per year. Such historical data have been used to show the pattern of gradual uplift centred upon the thickest part of the ancient ice cap (Figure 2.3b). Most recently, similar rates have been confirmed using satellite-based surveying techniques (e.g. GPS) and the area in which the initial measurements were made has become a UNESCO World Natural Heritage Site. The response to the removal of the ice sheet continues today, a long time after the ice sheet melted, indicating that the asthenosphere is extremely viscous. None the less, these slow-motion adjustments of the Earth's surface demonstrate that the Earth's interior is mobile.

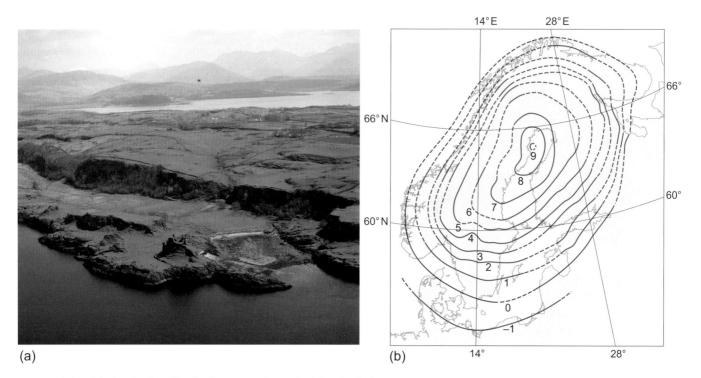

(a)   (b)

Figure 2.3   (a) Castle Coeffin (in foreground) on the island of Lismore, western Scotland, built on a wide platform of limestone and slate that extends inland to the foot of a line of former sea cliffs that are 5–15 m high. These cliffs mark the position of the shoreline before the land surface rose further out of the sea in response to unloading of an ice sheet. Reproduced with the permission of the British Geological Survey © NERC. All rights Reserved. (b) Map of the Scandinavian region showing contours of annual uplift rate in millimetres per year, based on data from 1892–1991. Contours are dashed where interpolated.

[1]Notice that this effect – an apparent decrease in ocean volume – is opposite to the rise in sea level expected from increasing the ocean volume by melting land ice.

2.3 Plate tectonics

The fluid nature of the asthenosphere allows the lithosphere to rise and fall in response to the influence of changing loads on the surface. As well as moving vertically, the lithosphere can also move horizontally, and this is responsible for many geological processes. The lithosphere is not an intact shell. Rather, it is divided into several large pieces called **plates** (Figure 2.4) consisting of crust and the immediately underlying lithospheric part of the mantle. Most plates contain both continental and oceanic crust. Each plate is moving relative to its neighbours, at typical speeds of a few cm y^{-1}. This happens because the asthenosphere is deformable, as you saw in the last section. **Plate tectonics** is the name given to the processes involved in the movement and interaction of the lithospheric plates.

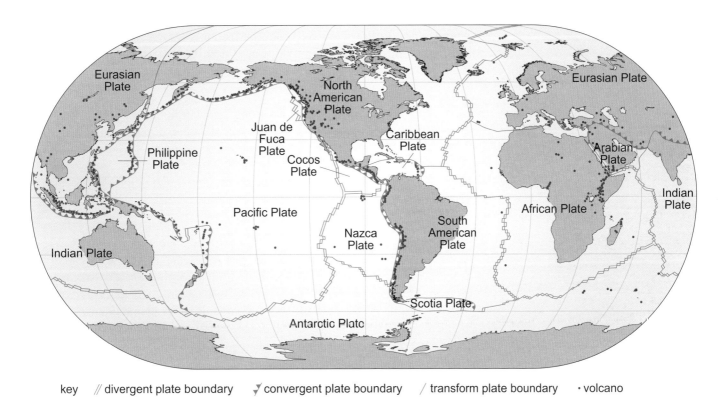

key   // divergent plate boundary   ⌐ convergent plate boundary   / transform plate boundary   • volcano

Figure 2.4   Map of the Earth's lithospheric plates and the locations of volcanoes. The three types of plate boundary identified in the key are discussed in the text. The triangles along the convergent plate boundaries are on the plate that is overriding its neighbour; the triangles point in the direction in which the subducting plate is moving relative to the overriding plate.

We will review plate tectonic processes very briefly by considering boundaries between adjacent plates in three circumstances, i.e. where the relative motion is convergent, divergent and parallel.

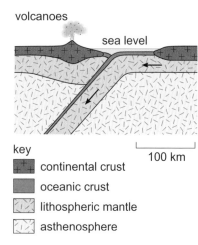

key

Figure 2.5   Cross-section through a subduction zone at a convergent plate boundary. The angle of descent shown here is fairly typical, but extreme cases dip as gently as 15° or as steeply as 80°.

2.3.1 Plates converging

Where two plates are moving towards each other, i.e. at a **convergent plate boundary**, one of them could dive down below the other, or the two plates could buckle and pile up on the surface. Which of these possibilities will happen?

■ Suppose the top of the colliding edge of one plate consists of continental crust, whereas the top of the colliding edge of the other plate is oceanic crust. Bearing in mind the relative densities of the two types of crust, which plate would be easiest to force downwards?

☐ Oceanic crust is denser than continental crust, so the edge of the plate carrying oceanic crust would tend to sink below the continental plate edge.

This situation is illustrated in Figure 2.5, which shows a convergent plate boundary, sometimes referred to as a 'destructive plate boundary' because the downgoing plate is 'destroyed' (although, in fact, it is recycled in complex ways through the mantle). Another term that is used is **subduction zone** because the act of one plate diving below another is termed **subduction**.

Where the subducting and overriding plates grind together, earthquakes are triggered. By determining the locations of these earthquakes in three dimensions, the downgoing plate can usually be traced to depths of several hundred kilometres. Subduction zones are also characterised by volcanoes on the overriding plate (Figure 2.4), a classic example being the line of volcanoes along the Andes Mountains in South America, located a few hundred kilometres east of the plate boundary.

When the colliding edges of both plates carry oceanic crust, it is not so obvious which one will be subducted. Generally speaking, the plate with the oldest, and therefore coldest and densest, oceanic crust is the one to sink below the other plate. When a subduction zone occurs within an ocean, a row of volcanic islands erupt through the overriding plate, just as at ocean–continent subduction zones. The volcanic **island arcs** of the western Pacific Ocean and the Caribbean are good examples. If volcanism and intrusion of igneous rocks at depth are prolonged, then the belt of affected crust can be so thoroughly modified by the addition of igneous material that it becomes more continental than oceanic in character, and indeed this is how continental crust is thought to have formed originally (and may still be forming today).

Subduction-related volcanoes occur only on the overriding plate, and tend to be found about 100 km above the subducting plate. Where the two plates meet, the downward bend of the subducting oceanic plate means that the sea floor at the plate boundary is particularly deep. The plate boundary is marked by an ocean **trench** that can be as deep as 8 km below sea level.

When two plates are converging, the main driving force is believed to be the weight of the relatively dense oceanic lithosphere pulling it below the less-dense continental lithosphere (or the island arc) of the overriding plate. Once this situation is established, there appears to be nothing that can stop it continuing until something really drastic happens. In Figure 2.5, the subducting plate is shown as part oceanic and part continental, with the oceanic part at its leading edge being subducted. What will happen if the clock is run forwards?

Question 2.3

In Figure 2.5, there is about 80 km of oceanic lithosphere left to be subducted. If the plate on the right is moving towards the plate on the left at a rate of 1 cm y⁻¹, how long will it be before all the oceanic lithosphere has been subducted?

The situation after this time is shown in Figure 2.6a. The weight of the subducting oceanic part of the plate will continue to drive the convergent motion, but now it has to pull down continental crust that has arrived at the subduction zone. This is less dense and thus more buoyant than oceanic crust, so it cannot be pulled down far and eventually the colliding continents lock up and subduction grinds to a halt (Figure 2.6b). The result at the surface is a **suture zone** marked by a mountain belt and a few slivers of oceanic crust and upper mantle that have avoided subduction.

Compressional deformation of the leading edge of the overriding continent and emplacement of the partly subducted continental edge of the other plate produce a mountainous region of greatly thickened continental crust. One region of the Earth where this has happened relatively recently (and is still happening) is where the northwards moving Indian continent has collided with the rest of the Asian continent, pushing up the Himalayan mountain belt. You will study the consequences of this and other continent–continent collisions in Chapter 10.

The different behaviours of oceanic and continental lithosphere at convergent plate boundaries – oceanic lithosphere is readily subducted whereas continental lithosphere is not – have a fundamental impact on the characteristics of the Earth's surface geology. Thus, most of the continental crust is very old, going back several billion years, and has a complex history, whereas the oldest oceanic crust is only about 200 million years old.

Figure 2.6b shows that the crust beneath a newly formed mountain belt is much thicker than normal. As noted in Section 2.2.1, the crust beneath the Himalaya and Tibetan plateau is up to 90 km thick in places, which is more than twice the thickness of the continental crust in tectonically stable regions. Intriguingly, the difference in thickness is much greater than the difference in elevation: the tip of Mount Everest is only about 9 km above sea level. This implies that the excessive crustal thickness must be accommodated largely by a downward deflection of the Moho (crust/mantle interface) rather than by the creation of mountains many tens of kilometres high. The situation looks more like Figure 2.7a than Figure 2.7b.

The reason that Figure 2.7b is unrealistic has to do with the relative densities of the crust and mantle, together with the fluid nature of the asthenosphere, which allows the mantle to flow away from areas of heavily loaded (thickened or ice-laden) crust. Blocks of crust with different thickness or density will 'float' in the mantle at their level of neutral buoyancy, like a ship, iceberg, or rubber duck floating in water. This principle is known as **isostasy**, and blocks of crust that have neutral buoyancy are said to be in **isostatic equilibrium**. As the crust has a lower density than the mantle, crust floats in the manner of an iceberg in water, so at isostatic equilibrium, high mountains are underlain by a very deep 'keel' within the mantle (Figure 2.7a).

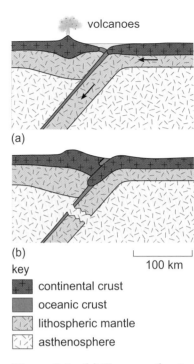

Figure 2.6   (a) Cross-section to show the outcome of continuing subduction for tens of millions of years beyond the stage shown in Figure 2.5. (b) A few tens of millions of years later still, a continent–continent collision has caused subduction to cease, and the slab of subducted oceanic lithosphere may break free, as illustrated here, and sink deep into the mantle. Note that volcanism has stopped.

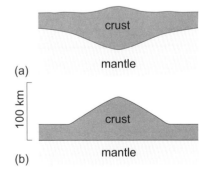

Figure 2.7   Cross-sections showing: (a) how thickened crust has a deep keel extending into the mantle; (b) the unrealistic topography required if the Moho were horizontal.

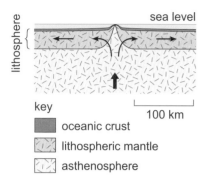

Figure 2.8   Cross-section through a divergent (constructive) plate boundary, drawn to the same scale as Figures 2.5 and 2.6.

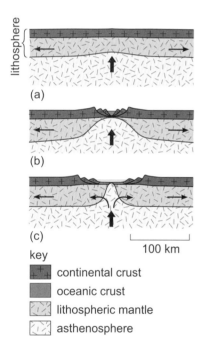

Figure 2.9   Time-series of cross-sections at the same scale as Figures 2.5, 2.6 and 2.8, illustrating divergent motion within a continent. In (a) and (b), upwelling within the asthenosphere causes heating and thinning of the continental lithosphere. By stage (c), the continent has rifted apart and a new ocean (e.g. the Red Sea, between Arabia and Africa) is beginning to open between them.

Activity 2.1   Isostasy

In this activity, you will investigate the connection between surface elevation and crustal thickness.

2.3.2 Plates moving apart

When two plates diverge, new oceanic lithosphere is created between them in a process aptly named **sea-floor spreading**, as shown in Figure 2.8. Such a plate boundary is known as a **divergent plate boundary** although it can also be known as a constructive plate boundary because new oceanic lithosphere is constructed there.

In sea-floor spreading, the divergent plate motion draws the asthenosphere upwards along the length of the plate boundary. The rising asthenosphere begins to melt (for reasons that will be explained in Chapter 5), releasing basaltic magma that escapes upwards and erupts or intrudes to create new oceanic crust. The upwelling asthenospheric mantle loses heat as it approaches the surface, becoming strong enough to become part of the lithosphere on either side of the plate boundary. The newly formed lithosphere continues to cool and become denser as it ages and moves farther from the plate boundary. The youngest oceanic lithosphere is the warmest and therefore has the greatest buoyancy, so it floats higher than older oceanic lithosphere – a result of isostatic equilibrium. Divergent plate boundaries therefore form broad topographic highs on the ocean floor. Although not all of them are midway across an ocean, divergent plate boundaries are often referred to as **mid-ocean ridges**. Typically, the crest of the ridge is 2–3 km below sea level, whereas the older, colder oceanic lithosphere far from the ridge has subsided isostatically to an average depth of 4–5 km. One of the best-known mid-ocean ridges is the Mid-Atlantic Ridge, which snakes its way from north to south through the Atlantic Ocean (Figure 2.4). Sea-floor spreading at the Mid-Atlantic Ridge drives North and South America away from Europe and Africa at a rate of a few centimetres per year.

Because divergent motion creates oceanic lithosphere, it should not be surprising that constructive plate boundaries are found today only within oceans (Figure 2.4). However, it sometimes happens that divergent motion is initiated within a continent, leading to the events shown in Figure 2.9.

The whole region in Figure 2.9a belongs to a single plate, but by the stage shown in Figure 2.9c the two halves are separated by a divergent plate boundary like that in Figure 2.8 and so belong to two separate plates. What was initially a single continent has rifted into two parts, each joined to new oceanic lithosphere at newly formed continental margins. The time interval between stages (a) and (c) is of the order of tens of millions of years. The faults shown in the process of formation in Figure 2.9b are normal faults, and are common in continental margins formed by rifting. You will learn more about this type of situation in Chapter 10.

2.3.3 Oceans opening and closing

Elements of the processes shown in Figures 2.5, 2.6, 2.8 and 2.9 are combined in Figure 2.10 to illustrate the life cycle of an ocean basin. In Figure 2.10a, a young ocean has opened and is becoming wider. In stage (b), a subduction zone has formed near one side of the ocean. Whether or not the ocean continues to widen depends on the relative rates of subduction and sea-floor spreading. In this example, subduction is outstripping sea-floor spreading so that by the time stage (c) is reached, the divergent plate boundary lies close to the subduction zone and will soon be lost. This has happened by stage (d); the ocean basin now lacks a constructive plate boundary, and will eventually be destroyed in a continent–continent collision. The whole cycle takes on average about 400–500 million years.

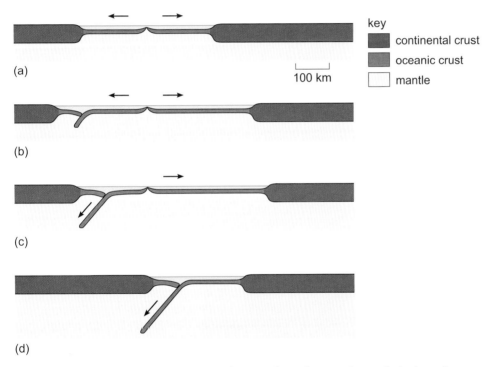

Figure 2.10   Time-series of cross-sections to show the opening and closing of an ocean. The scale is reduced compared to that of Figures 2.5, 2.6, 2.8 and 2.9, and the mantle lithosphere is not distinguished from the asthenosphere. See text for discussion.

The present northern Atlantic Ocean is at the equivalent of stage (a), having a mid-ocean ridge but no destructive plate boundaries. The central Atlantic Ocean may be likened to stage (b), with subduction occurring beneath the Caribbean Plate. The Pacific Ocean is most similar to stage (c). Its constructive plate boundary is displaced to the east side of the ocean, as you can see in Figure 2.4, but the situation is complicated because the Pacific Plate is being subducted on both sides of the ocean.

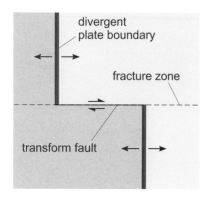

Figure 2.11   Sketch map showing how two plates (pale grey and pale blue) are separated by two divergent boundaries connected by a transform fault boundary. A fracture zone is the apparent continuation of a transform fault beyond the overlap, usually manifested by a scarp; however, the fracture zone is not a plate boundary and there is little or no motion across it.

2.3.4 Plates moving past each other

We may have given the impression that the relative motion between adjacent plates is always perpendicular to the boundary between plates. This is generally true in the case of divergent plate boundaries, but plates often converge at oblique angles. For example the Pacific Plate is approaching the North American Plate at a low angle along the Aleutian Trench (in the North Pacific), but the pattern of earthquakes and volcanism is little different from that at any other subduction zone with more 'head on' convergence.

However, there is a third type of plate boundary that we have yet to consider, which is where adjacent plates move at such a highly oblique angle that they slide past each other rather than form a subduction zone. In such a situation there is neither creation nor destruction of lithosphere. Examples abound at mid-ocean ridges, where divergent plate boundaries are offset by tens or hundreds of kilometres along **transform fault boundaries** (Figure 2.11). These boundaries (sometimes called conservative plate boundaries) are responsible for the rather jagged courses of mid-ocean ridges.

Transform fault plate boundaries can also occur within continental crust. The San Andreas Fault of California (Figure 2.12) is the most famous example and is responsible for several recent damaging earthquakes. Figure 2.4 also depicts the San Andreas Fault; it links the divergent plate boundary between the Pacific Plate and the Cocos Plate to the divergent plate boundary further north between the Pacific Plate and the Juan de Fuca Plate.

Figure 2.12   Oblique view, looking northeast, across the trace of the San Andreas Fault in southern California. This is a transform plate boundary between the North American Plate (San Bernardino Mountains in background) and the Pacific Plate (San Bernardino Basin, filled with sediments from surrounding highlands, in foreground).

2.4   The Earth's heat

Recall from Section 2.2.2 that the Earth's interior is hot. The sub-lithospheric mantle is so hot that it can deform slowly, with the result that the crust and lithosphere are able to move vertically and horizontally. Without these movements there would be little in the way of geological processes happening inside the Earth, which would be rather dull. Heat continually escapes from the interior. Within the mobile mantle, heat is transported by slow convection currents to the base of the lithosphere. Heat is then transferred to the surface by conduction through the lithosphere, by hot asthenospheric mantle approaching the surface at divergent plate boundaries and by active volcanoes. The total global rate of heat loss is about 4×10^{13} W. This works out at an average heat flow of about 0.08 W from every square metre (i.e. 0.08 W m^{-2}) of the Earth's surface, though it varies considerably from place to place.

W is the symbol for a watt, the SI unit of power. I watt = I joule per second (I W = I J s^{-1}).

The Earth's heat is relevant for understanding many of the processes that you will study later in this book, but where does this heat come from? Very approximately, and estimates vary, about 50% is reckoned to be primordial heat inherited from the time of the Earth's formation, and the remainder represents heat that is still being generated by the decay of radioactive elements. All radioactive decay produces heat, which is described as **radiogenic heating**, but there are only three elements today whose decay produces significant amounts of heat within the Earth. These are thorium, uranium and potassium.

Thorium (Th) in the Earth consists almost entirely of a single isotope, ^{232}Th. This symbol means the element thorium has 232 heavy particles (protons and neutrons) in its nucleus. It is usually referred to in speech as 'thorium-two-three-two'. ^{232}Th decays with a **half-life** of 14 billion years (i.e. 14×10^9 years). This is too slow a rate to be of much use in radiometric dating, but nevertheless this isotope acts as a significant heat source. Natural uranium (U) has two radioactive isotopes, ^{235}U and ^{238}U, whose half-lives are suitable for use in radiometric dating. Because ^{235}U has a shorter half-life than ^{238}U, the proportion of ^{235}U relative to ^{238}U has been declining ever since the Earth's origin and today only 0.7% of uranium is ^{235}U. Most potassium consists of the stable isotope ^{39}K, but 0.01% of it today is the radioactive isotope ^{40}K, whose decay to argon is much used in radiometric dating. Table 2.3 lists the relevant isotopes of these **heat-producing elements** and shows the rate of heat generation per kilogram of each isotope.

Table 2.3 Half-lives and rates of heat generation for isotopes of the important heat-producing elements in the Earth today.

Isotope	Half-life /10^9 y	Heat generation / µW kg^{-1}*
^{40}K	1.277	29.17
^{232}Th	14.05	26.38
^{235}U	0.7038	568.7
^{238}U	4.468	94.65

* 1 µW = 10^{-6} W

Table 2.3 shows that ^{40}K and ^{232}Th produce roughly similar amounts of heat per kilogram of isotope. However, given that only 0.01% of all K is ^{40}K, you might be surprised that ^{40}K is an important heat source. The explanation is that K is several orders of magnitude more abundant in the Earth than either Th or U, so even though only a tiny amount of potassium is ^{40}K, there is still enough ^{40}K to be significant.

The absolute and relative abundances of the heat-producing elements in different kinds of rock are crucial in controlling where in the crust and mantle most of the heat is produced. You can see from Table 2.2 that K is much more abundant in continental crust than in oceanic crust, which is itself richer in K than the mantle. Similarly, the abundances of U and Th, and therefore their rates of radiogenic heat production, are much greater in granite (which is a felsic rock type) than in basalt (mafic), and even less in peridotite (ultramafic). The rate of radiogenic heat production in intermediate rocks falls somewhere between the values for granite and basalt.

Radiogenic heating (per kilogram of rock) is greatest in continental crust and is particularly great in the upper part of the continental crust where most of the felsic rocks occur. The greatest rates of heat generation beneath the surface will therefore be where the continental crust is thickest, and that will be in regions of continental collision. You will meet some of the consequences of this increased heat generation in Chapter 10.

In global terms, taking into account the fact that the mass of the mantle is two orders of magnitude greater than that of the crust, the total radiogenic heat generated in the whole mantle works out at approximately equal to that generated in the whole crust. Primordial heat and radiogenic heat maintain the Earth in a state that allows the interior of the planet to be mobile. It also allows localised melting within the Earth, which brings us to the study of volcanoes and igneous processes in the next chapter.

2.5   Summary of Chapter 2

1   The Earth is layered compositionally into inner core, outer core, mantle and crust. The core is made from metal (dominantly iron) and the mantle from ultramafic rock with the composition of peridotite. There are two kinds of crust: oceanic and continental. Oceanic crust is basaltic in composition and has a higher density than continental crust, which is compositionally complex but on average is intermediate in composition. Oceanic crust is about 7 km thick; continental crust is about 38 km thick but can reach 90 km thick under some mountain belts.

2   The outer part of the Earth can be divided into two mechanical layers: the rigid lithosphere, consisting of the crust and the uppermost mantle, and the weak asthenosphere underlying it. The lithosphere is typically about 100 km thick, but is thinner beneath parts of the ocean and can exceed 200 km beneath ancient continents.

3   Below the lithosphere, mantle convection transports heat outwards towards the Earth's surface.

4   The weakness of the asthenosphere enables (i) the crust and lithosphere to reach isostatic equilibrium, and (ii) lithospheric plates to move over the asthenosphere.

5   Oceanic lithosphere is created at divergent plate boundaries by sea-floor spreading and 'destroyed' at convergent plate boundaries by subduction. Most of the continental crust is much older than any surviving oceanic crust. Collision of continental lithospheres causes crustal thickening and forms mountain belts with deep crustal roots.

6   The heat escaping from the Earth's interior is derived from primordial heat and heat being generated by decay of the radioactive isotopes of U, Th and K. These heat-producing isotopes are particularly concentrated in the continental crust.

2.6   Objectives for Chapter 2

Now you have completed this chapter, you should be able to:

2.1 Describe the Earth's internal structure, and explain how this is known.

2.2 Describe the rudiments of plate tectonics, giving examples of different kinds of plate boundary.

2.3 Make simple calculations to solve problems related to isostatic equilibrium.

2.4 Describe how heat is generated within the Earth and how this varies between different rock types.

Now try the following questions to test your understanding of Chapter 2.

Question 2.4

Describe what each of the following reveals about the Earth's interior:

(a) the ratio between the Earth's mass and volume

(b) the way in which seismic waves pass through it

(c) the Earth's magnetic field

(d) the rise of ancient shorelines above present-day sea level in areas that were covered by thick ice caps.

Question 2.5

(a) Imagine that the crust at the top of column C in Figure 2.13 is subjected to erosion. In what way will the depth of the base of the crust in column C change as the top of column C is eroded, and why?

(b) Suppose the sediment derived by erosion from the top of column C is deposited on the top of column D. What will happen to the crust in column D?

(c) Where is the actual flow taking place that allows the base of the crust to change its depth?

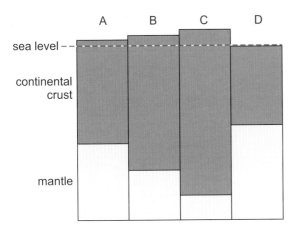

Figure 2.13   Cross-section dividing the crust into blocks A–D resting in isostatic equilibrium upon the mantle.

Question 2.6

Briefly describe two independent lines of evidence that indicate the presence of a weak asthenosphere immediately below the lithosphere.

Question 2.7

Is the following statement true or false? 'The whole of the Atlantic Ocean is underlain by a single tectonic plate.' Give your reasons.

Question 2.8

In which of the following settings would you expect to find most heat per square metre reaching the Earth's surface by conduction: (a) on old ocean floor; (b) on a region of continental crust slightly above sea level; (c) in a mountain belt produced by continent–continent collision? Give your reasons.

Chapter 3 Volcanic rocks

Active volcanoes are one of the places on Earth where geological processes can actually be observed while they are happening, so studies of volcanic eruptions give geologists insights into how volcanic rocks in the geological record can be interpreted. In studying the formation of igneous rocks from cooling magma, it is also useful to start by elaborating on what magma actually consists of and learning about some of its physical properties.

3.1 Magma

Magma is molten rock, and is made up of a liquid (sometimes referred to as 'melt') usually containing some crystals and/or some gas bubbles suspended in the liquid.

There are three important physical properties that influence how magmas behave and, therefore, how igneous rocks form. These are the temperature, density and viscosity of the magma.

The crystallisation temperatures of minerals found in igneous rocks (Book 1, Figure 6.6) help in answering the question 'How hot is magma?'. Minerals that are rich in Mg or Ca and low in Si tend to have high melting temperatures, whereas minerals that are poor in Mg or Ca but rich in Si tend to have lower melting temperatures. The chemical composition of a magma will therefore determine the temperatures at which it is molten. Thus, mafic magmas can be as hot as 1200 °C, whereas some felsic magmas can be no more than 700 °C. Intermediate magmas have temperatures between these extremes although, as with chemical composition, there is a continuous range between the extremes.

An important property of magmas is that they tend to be slightly less dense than their solid equivalent. This means that magmas have a tendency to rise buoyantly through the mantle and crust. How freely magma is able to move, both at depth and when erupted at the surface, depends on its **viscosity**. This is a measure of its resistance to flow, and is measured in units of kg m^{-1} s^{-1} – more commonly expressed as pascal seconds (Pa s). (The pascal is the SI unit of pressure and 1 Pa = 1 kg m^{-1} s^{-2}, so 1 Pa s = 1 kg m^{-1} s^{-1}.) The viscosities of mafic, intermediate and felsic magmas are typically of the order of 10^2, 10^3 and 10^5 Pa s, respectively (for comparison, the viscosities of water and golden syrup are about 10^{-3} and 10^2 Pa s). This range reflects the increase in viscosity due to (i) decreasing temperature and (ii) increasing SiO$_2$ content of the magma. You should also be aware that the proportion of crystals within the magma will affect the viscosity.

■ All else being equal, would you expect an increase in the abundance of crystals to impede the ability of a magma to flow?

☐ Adding a lot of solid particles, such as crystals, will interfere with the ability of the liquid to flow, so crystallinity increases the viscosity of a magma.

Setting aside the influence of temperature and crystal content, to understand how the purely compositional influence on viscosity arises requires us to consider magma on the atomic scale. Recall that rock-forming silicate minerals are based on the SiO_4 tetrahedron arranged with metal cations in regular repetitive units. Magmas and molten minerals do not have this regular crystalline structure. They consist of a mixture of negatively charged SiO_4 tetrahedra that occur as isolated units, or linked together (polymerised) in chains or frameworks, together with positively charged metal ions that tend to interrupt the chains. The lack of more than purely local order in the structure of such a mixture allows it to behave as a liquid. The more felsic the magma is, the more abundant and longer the chains of SiO_4 tetrahedra.

■ Do you think that the presence of silicate chains and frameworks will increase or decrease the viscosity?

☐ Long chains hinder the flow of a magma, and this is why felsic magmas are more viscous than mafic magmas.

In the same way that positively charged metal ions disrupt the chains of SiO_4 tetrahedra, gases that are dissolved in magma also cause a reduction in the viscosity of magma. Such dissolved gases are known as **volatiles** and water (H_2O) is the most abundant, but carbon dioxide (CO_2), sulfur (S), chlorine (Cl) and fluorine (F) are also found dissolved in magmas. The volatiles occur as isolated molecules dispersed among the silica chains and also as anions that break up the silicate chains, so volatiles play a role in lowering the viscosity of magma.

The amount of volatiles that a magma can hold increases with pressure. However, if the pressure falls, some of the volatiles will begin to **exsolve** (i.e. separate out as a distinct phase, forming gas bubbles).

■ Suggest a natural situation in which the pressure experienced by a body of magma would fall.

☐ The most obvious case is when magma is rising upwards through the Earth.

This tendency for volatiles to exsolve from magma as it rises to shallower depths is the main driving force behind explosive volcanic eruptions. Magmas that are poor in volatiles tend to produce less-violent eruptions in which lava flows over the ground; these are known as **effusive eruptions**.

3.2   Lava flows on land

Magma flowing at the surface is described as **lava**, and the same term is used to describe the rock after it has solidified. Whether magma erupts from a volcano at an isolated point, or **vent**, or from a linear crack known as an **eruptive fissure**, the lava flows can travel for hundreds of metres to many kilometres. The longest known lava flow of relatively recent geological age is the 120 km-long Toomba flow, erupted in Australia about 13 000 years ago. But even this exceptional flow is dwarfed by others, possibly as long as 1000 km, that have been recognised by geological mapping of Cretaceous to Palaeocene volcanic rocks in India.

The form taken by a lava flow depends very much on how the lava actually moves. Factors controlling this movement include the steepness of the slope upon which the flow is emplaced, the rate at which lava is supplied at the source (the **effusion rate**), and the cooling rate.

■ There is another important factor. What do you think it is?

☐ The form of a lava flow must depend very strongly on how freely the lava is able to flow, in other words on its viscosity.

Viscosity decreases with increasing volatile content and temperature, and increases with the proportion of crystals contained in the melt. However, as you saw in Section 3.1, bulk composition of the magma is also an important factor. Lower silica contents give magma very much lower viscosity, so mafic lava can spread more thinly and flow farther down gentler slopes than intermediate and felsic lava.

Some of the Earth's most active basaltic volcanoes are to be found in Iceland, Hawaii and Réunion Island (in the Indian Ocean), and they provide excellent 'natural laboratories' for studying lava flows. Despite all these volcanoes being basaltic, a surprising range of surface morphologies can be seen on these lava flows. At one extreme, the surface takes the form of rough clinkery blocks, whereas at the other it consists of a smooth surface that may be wrinkled and draped to give the appearance of coiled rope (Figure 3.1). These two surface textures are known to volcanologists by the Hawaiian terms **a'a** (the pronunciation of which – ah-ah – is said to resemble the cries of pain uttered when walking barefoot over such a loose and jagged surface) and **pahoehoe** (pronounced 'pa-hoey-hoey' or 'pa-hoy-hoy').

Figure 3.1   An a'a flow surface (left) partially overridden by a subsequent ropy pahoehoe flow (right), Hawaii. The flows are identical in composition, but the pahoehoe looks paler because of its shiny fresh glassy surface.

The factors controlling the development of lava surface morphology have been deduced from observing active lava flows and from laboratory experiments using hot wax to simulate lava. These show that pahoehoe surfaces can form only when the rate of flow is slow enough to allow the surface to chill and produce a solid but pliable crust a few millimetres thick over most of the flow surface, whereas a'a texture originates when the flow is too rapid to allow a coherent crust to survive. A range of morphologies is recognised depending on flow speed.

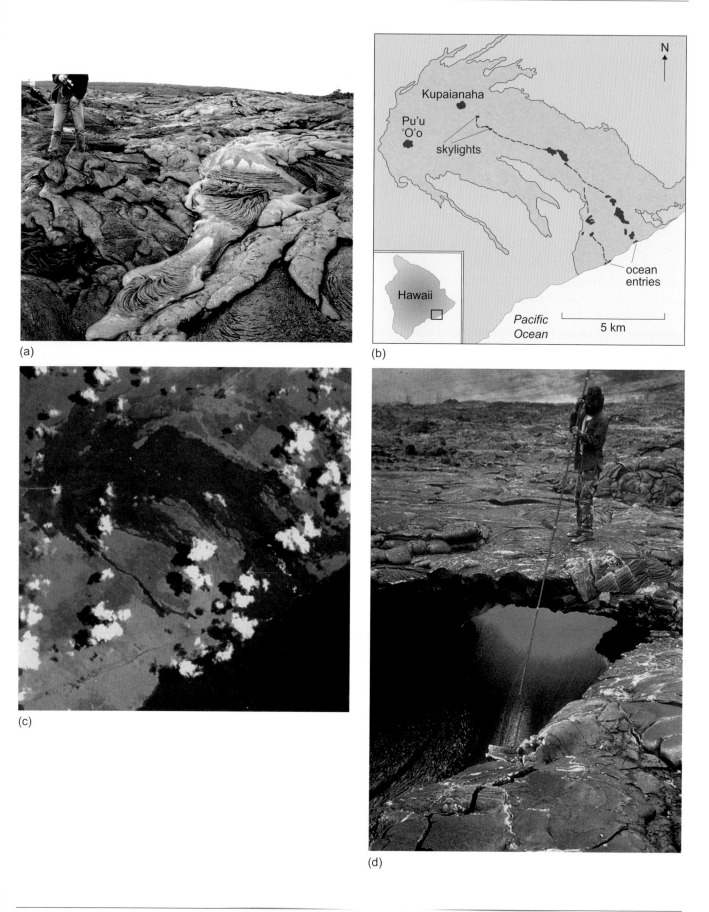

(a)

(b)

(c)

(d)

Figure 3.2 illustrates the development of pahoehoe lava, using examples from Hawaii. At the front of slowly moving pahoehoe lava (Figure 3.2a) the surface crust is cooling and thickening all the time, and manages to retain lava that is flowing into it from farther upstream. Instead of the lava constantly moving forwards, the new lava can be accommodated by the flow becoming deeper rather than longer. It gradually inflates to form a metre-scale lobe or toe until eventually the lava crust can stretch no further and it ruptures, releasing a new toe of incandescent lava. The process repeats time and time again, developing into an extensive area of active and inactive lobes known as a **flow field** (Figure 3.2b), which grows as lobes advance in seemingly random patterns at the flow front. Magma is supplied through a sinuous arterial network linking the vent to the flow front, although this feeding system may only be visible by using thermal imaging devices (Figure 3.2c) or by spotting the locations of **skylights** in the roofs of subsurface **lava tubes** (Figure 3.2d).

At faster flow rates and on steep slopes, lava is less able to spread out sideways so the central part of the flow moves faster than its sides. This soon develops into a channel of flowing lava hemmed in by a low wall (known as a **levée**) of solidified lava or rubble on either side (Figure 3.3). The relatively rapid flow within the lava can continually break the developing lava crust, resulting in a ragged a'a surface.

On close inspection, basaltic lava is often found to contain bubbles, usually a few millimetres across (Figure 3.4). These are known as **vesicles**, and are created by the **exsolution** of volatiles from the magma at low pressure. The gas bubbles were unable to escape before the lava solidified, so the end result is a vesicular rock. Vesicles may be particularly abundant in a'a lava, and contribute to the clinkery appearance of its surface.

Lavas of intermediate and felsic composition are too viscous to develop pahoehoe textures; instead their surfaces break into slabs or blocks – like a larger scale version of a'a, but with fractured, less clinkery, surfaces (Figure 3.5). These are called **blocky lava** and, generally speaking, the more viscous the flow, the larger the blocks.

Figure 3.3   Lava channels and their levées formed during the 1984 eruption of Mauna Loa volcano, Hawaii. The channel width varied from 20 to 50 m.

Figure 3.4   Cross-section through a fresh sample of pahoehoe lava (Hawaii), showing a particularly high concentration of vesicles. In this specimen, the vesicles are almost spherical, a characteristic of pahoehoe.

Figure 3.2   Pahoehoe lava. (a) Pahoehoe lobes formed by a series of breakouts, advancing over older lobes. (b) Map of the lava flow field produced from Kilauea volcano, Hawaii, between 1983 and 1991. Main vents (Pu'u 'O'o and Kupaianaha), skylights and lava tubes active on 23 July 1991 are shown in red. (c) Thermal satellite image of the flow field shown in (b), revealing the active parts of the field against a green background of vegetation. The red spot near the left edge is an active vent from which fumes drift to the south. Other red areas are breaks in the lava tubes that transport lava from Kupaianaha to the coast. (d) Lava flowing through a lava tube revealed by a skylight within a pahoehoe lava field characterised by lobes with smooth and ropy surfaces.

Figure 3.5   The blocky surface of a rhyolite lava flow (Obsidian Flow, Inyo Domes, California). Note person for scale (left of centre).

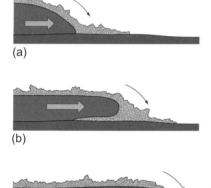

Figure 3.6   Schematic cross-section through a blocky or a'a lava flow, advancing from left to right. Such flows are typically from one metre to several metres in thickness, but can be much thicker. The interior of the flow is molten (red), but the surface is rubbly (grey). In (a), rubble tumbles from the front of the flow, and is progressively overridden (while fresh rubble continues to be shed at the front) in stages (b) and (c).

A'a flows and blocky lava flows usually form channels confined between rubbly or blocky levées and they move forwards in the way illustrated in Figure 3.6. Rubbly material from the flow top tumbles down the flow front and is overridden in a similar way to the rolling caterpillar track on a bulldozer. When a flow that has been emplaced in this manner is seen in cross-section its interior is usually solid (though vesicular), whereas both its upper and lower surfaces are rubbly.

When lava with an extremely high viscosity is erupted it cannot flow far and, therefore, builds up over the vent as a large mound, or **lava dome**, which can be 100 m or more in thickness. Lava domes and short thick lava flows are characteristic of felsic magma such as rhyolite and crystal-rich intermediate magma such as andesite and dacite (a rock type that covers the compositional range between andesite and rhyolite). Examples of these lava domes are illustrated in Figure 3.7.

Question 3.1

Figure 3.8 shows a graph of the lengths of a'a lava flows and their average effusion rates.

(a) Write a few sentences describing the relationship shown.

(b) Suggest a reason for the relationship you described in part (a).

(c) How useful might this graph be if you lived close to a volcano that frequently erupted lava flows?

Figure 3.7 Lava domes. (a) Dome of Soufrière Hills volcano, Montserrat, West Indies on 12 April 1996, composed of crystal-rich andesite. The spine of lava jutting out of the debris-covered dome is about 40 m high. (b) Little Glass Mountain, northern California, composed of crystal-poor rhyolite that erupted about 1000 years ago. The large volcano Mt Shasta is on the skyline. (c) Chao lava dome (in centre of lower left quadrant) in Northern Chile, is over 300 m high and 14 km long and is composed of crystal-rich dacite. It erupted within an extensive area of conical andesite volcanoes, as shown in this false-colour image generated from Landsat data. Two smaller domes are visible above and below the centre of the image, and a third is in the extreme lower right corner.

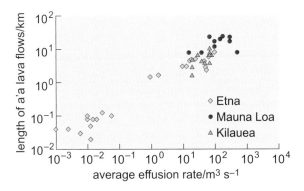

Figure 3.8 A graph of a'a lava flow length plotted against effusion rate. Data are shown for three volcanoes: Mauna Loa and Kilauea in Hawaii, and Etna in Sicily.

3.3 Lava flows under water

When basalt is erupted under water, it can develop a flow morphology known as **pillow lava**, a name that reflects the fact that the lava looks like an expanse of pillows piled up on top of each other (Figure 3.9). Each pillow has a glassy margin a centimetre or more thick and a dense more crystalline interior. Radial cooling cracks are sometimes present.

Figure 3.9   Pillow lava, which forms only when basalt is erupted under water. Notice how the undersides of later pillows have moulded themselves to fill the cuspate cavities between the bulbous tops of underlying pillows. This indicates that the rocks are still in their original orientation, even though they have been thrust from the ocean floor onto a continent (Oman, Arabia). Note geological hammer for scale.

■   What factor could produce the special characteristics of basalt pillow lava under water?

☐   Slopes and effusion rates under water can be the same as on land, and viscosity must be identical because this question relates only to basalts. However, under water the rate at which the surface of a lava flow is cooled could be very different to the cooling rate on land.

Water both absorbs and transports heat much more efficiently than air, so upon contact with water the skin of a lava flow is cooled extremely rapidly. Observations of submarine lava flows off the coast of Hawaii and at divergent plate boundaries have shown (Figure 3.10) that when the internal pressure of the lava ruptures the skin, lava flows out and quickly chills on the surface, only to burst and extrude another small lobe. Over time, a single submarine eruption produces thousands of pillow- or bolster-like lobes that are about 0.2–2 m in diameter, similar to those in Figure 3.9.

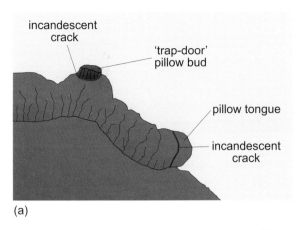
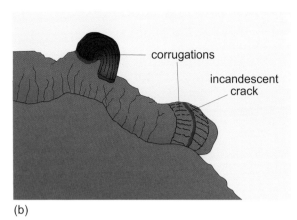

(a)

(b)

Figure 3.10 Stages in the development of pillow lava. (a) A pillow tongue advances over the sea floor, being fed from a source to the left. Incandescent cracks in its skin open up as the pillow is inflated by the continued injection of lava. (b) As the front of the pillow moves forwards, an incandescent crack opens, while at the same time a new pillow begins to form where lava escapes from a 'trap-door' crack in the roof of the original pillow.

Because the skin of a submarine lava flow is chilled so rapidly, it tends to flake away even as it forms, producing glassy fragments described as **hyaloclastite** (meaning 'broken glass'). These fragments may collect in the spaces between pillows, where they weather rapidly to form clay minerals. In more extreme cases, an entire flow may break into hyaloclastite fragments to give a hyaloclastite rock.

Pillow lavas characterise many eruptions of basalt under water, but morphologies similar to pahoehoe, and even a'a, are formed if the effusion rate is high enough. On the other hand, intermediate and felsic lavas are usually too viscous for their morphology to be controlled by the glassy crust produced by eruption under water, so they tend not to form pillows. The main effect of water in these cases is to encourage fragmentation, which can reduce the flow to a pile of shattered rubble. Although produced non-violently, hyaloclastite is fragmentary and so can be regarded as a pyroclastic rock. You will look at more conventional pyroclastic rocks and the eruptions that produce them in the next section.

3.4 Pyroclastic eruptions

You saw earlier how volatiles coming out of solution produce vesicles within an erupting lava. If the volume of these bubbles exceeds about 80% of the total volume, the rising magma will fragment explosively and be shot out of the vent of the volcano. This is the most dramatic way of creating a pyroclastic rock. Individual rock fragments (clasts) produced in these eruptions are known as **pyroclasts**.

■ Which two factors are likely to cause an eruption to be explosive rather than effusive?

☐ The main factors are the amount of volatiles available in the original magma, and the ease with which volatiles can escape as the magma approaches the surface.

The ease of escape of this gas depends on the viscosity of the magma. If the viscosity is low enough, bubbles can rise through the magma and escape, with the result that the volume of gas present at any instant may never reach 80%. Magma with low volatile content and low viscosity will therefore tend to produce effusive eruptions. On the other hand, explosive eruptions are favoured when the magma has high viscosity and high volatile content.

■ Which chemical type of magma has the highest viscosity?

☐ Felsic magmas are the most viscous (Section 3.1).

However, even basalts can, in some cases, contain enough volatiles to produce explosive eruptions. To understand why this can happen, you need to consider the conditions within the conduit supplying magma to the volcano's vent (Figure 3.11). Gas has a very low density, so buoyancy will always drive gas

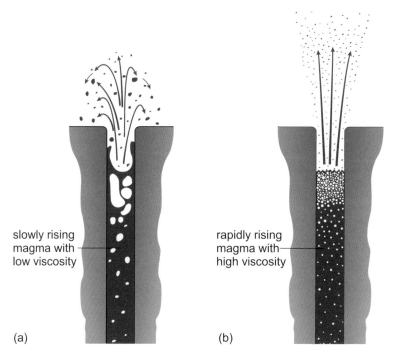

slowly rising magma with low viscosity

rapidly rising magma with high viscosity

(a)

(b)

Figure 3.11   The arrangement of gas bubbles in a magma conduit in two contrasting cases: (a) when low magma viscosity allows bubbles to coalesce and rise faster than the magma, globules of molten magma are ejected from the vent when the bubbles burst; (b) when the high viscosity of the magma causes gas bubbles to stay trapped within the rising melt, they build up enough pressure to unleash an explosive discharge of gas and pyroclasts.

bubbles to move upwards. In any liquid, including magma, small gas bubbles rise more slowly than large bubbles. So, in the situation where the magma has a low viscosity but is not rising very fast, gas bubbles can move upwards faster than the magma (Figure 3.11a). The rising gas bubbles can coalesce into larger bubbles, which move through the magma at an even faster speed, and by the time these bubbles reach the top of the magma they can be a metre or more across. They arrive at irregular intervals of a few tens of seconds and, on bursting, throw out a shower of molten lumps of magma. If the bubble bursts gently, the ejecta flop out around the rim of the conduit (Figure 3.12a). More violent bubble bursting throws ejecta to greater heights (Figure 3.12b), and this sort of episodic activity is described as **strombolian**, after the almost continuously active Italian volcano Stromboli, which exhibits this style of activity.

In more energetic basaltic eruptions, the expansion of vesicles in the magma conduit can cause the magma and bubbles to accelerate upwards together, forcing them out of the vent at speeds of the order of 100 m s^{-1} as a continuous fountain of incandescent lava fragments and gas (Figure 3.12c). Most of the material thrown up in the **fire fountain** falls to the ground nearby, although some finer particles may be carried aloft by the hot escaping gas or blown away by the wind. If the large clasts are still molten when they hit the ground, they are known as **spatter** (Figure 3.13a) and rapidly accumulating deposits of spatter can in some cases be hot enough to transform into flowing lava. However, if the clasts are chilled sufficiently in flight, they hit the ground as solid clinkery pieces called **scoria** (Figure 3.13b). The larger clasts flung out by an explosive eruption are described as **volcanic bombs**. In the case of a scoria-producing fire fountain, the

(a)   (b)   (c)

Figure 3.12   (a) Photograph taken at night a few seconds after a mild strombolian bubble-bursting event at the top of an active conduit in Masaya volcano, Nicaragua. The vent is about 5 m across and is nearly brimfull of molten magma. The bubble that has burst has thrown out molten spatter, which is already beginning to cool. (b) Night-time strombolian explosion at the summit of Stromboli, Italy, which launched incandescent spatter on ballistic paths, illuminating the crater walls (9 June 2008). (c) A 500 m-high fire fountain on Hawaii in 1985.

(a)

(b)

(c)

(d)

Figure 3.13 Examples of pyroclasts: (a) spatter fragments that have landed on a smooth, flat surface of lava, Kilauea volcano, Hawaii; (b) scoria; (c) a volcanic bomb with an aerodynamic shape, South Sister volcano, Oregon (lens cap gives scale); (d) pumice, about 10 cm across, Taupo volcano, New Zealand.

bombs chill less (because of their large size) and are typically soft and gooey so that they may take on streamlined aerodynamic shapes or splat against the ground on landing (Figure 3.13c).

In contrast to mafic magmas, intermediate and felsic magmas are too viscous to allow vesicles to merge or to rise independently of the magma. Instead, the vesicles remain small (less than a few mm across) and become more numerous as more and more new bubbles form in order to accommodate the decreasing volatile solubility as the pressure in the magma decreases. Once the rising magma reaches a state in which the vesicles are so tightly packed that they can no longer expand freely, the magma disintegrates explosively (Figure 3.11b). At this **fragmentation surface**, the erupting material changes from a viscous liquid containing gas bubbles into a decompressing jet of gas containing countless shattered fragments of frothy magma, or **pumice** (Figure 3.13d), with a large range of grain sizes (see Box 3.1). The expanding gas accelerates the mixture up the conduit, blasting out of the vent at several hundred metres per second. These eruptions are exceedingly violent.

Box 3.1 Notes on the terminology for pyroclasts

Explosive eruptions shatter the magma so effectively that particles of all sizes are ejected from the vent. In general, these fragments (or pyroclasts) are referred to as **tephra** (from the Greek word for 'ash'). Tephra can be fragments of solidified magma (referred to as **juvenile tephra**) or fragments of pre-existing rock ripped from the crater walls (**lithic fragments**). It is sometimes useful to be able to distinguish tephra according to grain size so, for reference, Table 3.1 shows the accepted grain size (i.e. particle size) terminology for tephra.

Table 3.1 Terminology for tephra according to grain size.

Grain size/mm	Pyroclastic fragments (tephra)
>256	coarse bombs and blocks
64–256	fine bombs and blocks
2–64	lapilli
$\frac{1}{16}$–2	coarse ash
$< \frac{1}{16}$	fine ash

Particularly in eruptions of felsic magma, many of the juvenile fragments in the lapilli size range are lumps of glassy froth called pumice. It is also worth remarking on the term **volcanic ash** (usually abbreviated to just ash) which describes fragments <2 mm in size produced by the disintegration of magma. Most ash particles are sharp glassy fragments representing the walls of bubbles, though some consist of isolated phenocrysts or are fragments of phenocrysts. The colour of ash – black (if mafic) or grey (if intermediate or felsic) – resembles the ashes of a fire, but neither volcanic ash nor cinders (a term sometimes used for larger clinkery fragments) are products of combustion.

3.4.1   Eruptions that produce fall deposits

Figure 6.9b of Book 1 shows an image of a pyroclastic eruption from the Soufrière Hills volcano on the island of Montserrat. In that eruption, a torrent of ash and gas cascaded down the volcano's slopes. Figure 3.14 shows another type of pyroclastic eruption. This time, a plume of ash and gas is rising high into the atmosphere. Eruptions of this type, with a grey column of ash rising kilometres to tens of kilometres above the vent, can last for hours to days and are known as **plinian eruptions**. The reason for the name is that the first detailed eyewitness account of such an eruption (that of Vesuvius in AD 79) was written by the 17-year-old Pliny the Younger. The eruption destroyed the cities of Pompeii and Herculaneum, and killed over 3500 people (including the Roman scholar and naturalist Pliny the Elder, uncle of Pliny the Younger).

A plinian **eruption column** has three parts (Figures 3.15). In the lower part of the column, the erupting mixture of gas and pyroclasts are shot upwards by the force of expanding gas in the conduit (as in Figure 3.11b). This is called the gas thrust region, and is equivalent to the entire eruption column of a fire fountain. However, in a plinian eruption column a lot of air is drawn in and heated by contact with the hot ejecta, causing the entrained air to expand. The overall mixture, despite the weight of the pyroclasts within it, becomes buoyant relative to the cold surrounding air so it rises convectively above the gas thrust region, entraining more air into the billowing column. This is called the convective ascent region of the eruption column. Eventually, the column reaches a height where it is neutrally buoyant in the atmosphere, and spreads out to form an umbrella cloud. Pyroclasts that fall out from the eruption column and the spreading umbrella cloud blanket the ground below. The resultant **fall deposit** drapes the landscape (Figure 3.16), just like a layer of snow.

Figure 3.14   A plinian eruption column rising above Pinatubo volcano, Philippines.

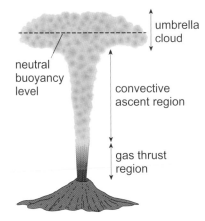

Figure 3.15   A plinian eruption column, showing the gas thrust region, the region of convective ascent, and the umbrella cloud.

Figure 3.16   A thick, white layer of pumice, and thinner, darker pyroclastic layers above it, exposed in a roadside cutting (Towada volcano, Japan). The parallel undulating lower and upper contacts indicate that the deposits mantled the pre-existing topography. They are therefore fall deposits.

Question 3.2

Look at Figure 3.14 and estimate the proportion of the total height of this particular eruption column that is occupied by the umbrella cloud.

Plinian fall deposits tend to become both thinner and finer-grained away from their source volcano. These patterns can be seen by visiting as many exposures of a given fall deposit as possible, measuring the thickness and plotting the results on a map (Figure 3.17a). Contours of equal deposit thickness (known as **isopachs**) can be drawn, which allow the volume of the deposit to be calculated. Likewise, measuring the average size of the largest pumice or lithic fragments reveals the sizes of the largest particles that were carried upwards by the eruption plume and how they were dispersed (Figure 3.17b and c). Contours of equal particle size are known as **isopleths**.

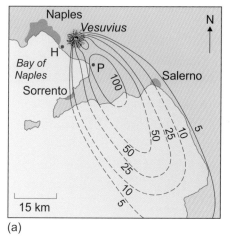

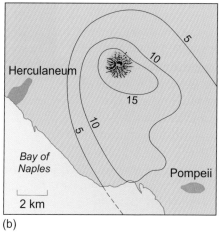

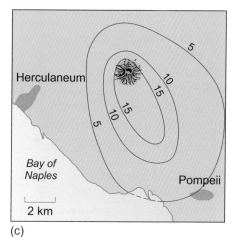

(a)   (b)   (c)

Figure 3.17   Maps of the lower part of the deposit produced in the AD 79 eruption of Vesuvius: (a) isopach map; (b) isopleth map of maximum pumice size; (c) isopleth map of maximum lithic size. All thicknesses are in centimetres. H = Herculaneum; P = Pompeii.

The characteristics of a fall deposit that can be mapped in the field reflect the dynamics of the eruption column itself. Thus, the largest pyroclasts are too heavy to be carried upwards by the convective ascent region, so they fall out rapidly, hitting the ground relatively close to the vent – although some bombs can travel several kilometres. Slightly smaller material falls from the sides of the convective ascent region, and is dispersed farther than the majority of the bombs. Tephra falls from the spreading umbrella cloud when the swirling motion within the cloud can no longer keep the particles aloft. This means that smaller particles tend to remain in the cloud for longer, so they are dispersed farther from the volcano than larger particles. The extent of the fall deposit therefore depends on how far the umbrella cloud spreads.

The thicknesses of tephra produced in a plinian eruption depend on the total amount (volume or mass) of material deposited and the energetics of the eruption column but there is an important external factor that can have a considerable influence on the *shape* of isopachs and isopleths.

■   What is this external factor?

☐   The wind, which controls the dispersal of the umbrella cloud.

When there is a strong wind at the time of the eruption, the umbrella region is blown off to one side (Figure 3.18) so the tephra is deposited asymmetrically. When the thickness of a fall deposit is contoured, the pattern reflects the wind direction during the eruption.

Figure 3.18   Photograph of the Kliuchevskoi (Kamchatka, Russia) eruption of 1 October 1994, showing downwind spreading of the plume in the upper atmosphere, photographed by Space Shuttle astronauts.

Question 3.3

Look at Figure 3.17. From which direction was the wind blowing during this eruption? Explain your answer.

Careful study and mapping of a fall deposit can tell us much about the eruption that produced it:

*   the wind direction at the time of eruption
*   the volume of magma erupted
*   the maximum height of the eruption column, calculated from the dispersal range of maximum particle sizes.

The details of the latter calculation are beyond the scope of this course, but we have already touched on the basic underlying principle – the higher the eruption column, the farther particles of a given size will be carried before falling to the ground. Furthermore, it has been discovered that the column height depends on the eruption rate (Figure 3.19).

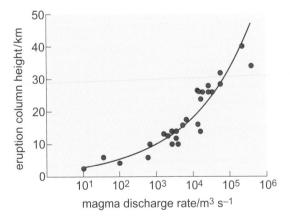

Figure 3.19   Plot of eruption column height against magma discharge rate. Discharge rate refers to the volume of dense (i.e. non-vesicular) magma erupted in unit time. The curved line is a theoretical model of a convective plume rising into the atmosphere.

Question 3.4

According to Figure 3.19, what was the eruption rate during the Kliuchevskoi eruption shown in Figure 3.14, given a column height of 12 km?

Figure 3.20   Part of a plinian fall deposit (Taupo volcano, New Zealand) showing a decrease in grain size around the level of the tape measure case and passing upwards to increased grain size. This reflects a period of decreased eruption rate part way through the eruption.

A common characteristic of volcanic eruptions is that the discharge rate of the magma changes over time. In the case of plinian eruptions, this would mean that not only will the average size of tephra in a fall deposit decrease away from the vent, it would also change upwards within the deposit at any given location (Figure 3.20).

3.4.2   Pyroclastic flows and their deposits

The previous section considered eruptions in which buoyant mixtures of gas and tephra rise into the atmosphere, and the deposits formed from this type of eruption. Here, we consider the opposite – eruptions in which the mixture of gas and tephra is denser than air. Under these circumstances, a **pyroclastic flow** of juvenile tephra, lithic fragments and gas sweeps across the ground away from the vent, with the material as a whole behaving as a dense fluid (it is said to be **fluidised**). A simple cross-section through a moving pyroclastic flow is given in Figure 3.21. The head of the flow moves forwards at speeds of tens to hundreds of metres per second (based on observations of historic examples such as those on Montserrat shown in Book 1, Figure 6.9b) and entrains air, further fluidising the material. This relatively dilute part of the pyroclastic flow is sometimes called a **surge** and can deposit tephra in rippled layers similar to those seen in some sandstones deposited by flowing water. The finest material in the pyroclastic flow tends to form an ash cloud, which rises from the head or body of the flow and is blown away. However, most of

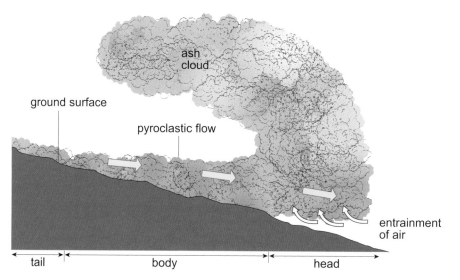

Figure 3.21   Cross-section through a pyroclastic flow (typically hundreds of metres to tens of kilometres in length), which is moving downslope from left to right.

the pyroclastic material is deposited on the ground from the body of the flow, which continues to advance until it loses the ability to entrain sufficient air to remain fluidised.

There are several ways in which a pyroclastic flow can be generated. One common way is when a plinian eruption column ceases to entrain enough air to keep it buoyant, such that it becomes unstable and instead of rising upwards it collapses to the ground. For a plinian eruption column to be unstable, the rate of magma discharge and the radius of the vent (which are related) must exceed a certain threshold, while the exit velocity at the vent and the volatile content of the magma (which also tend to be related) must be relatively low. The relationship is shown graphically in Figure 3.22. Study this figure, and then attempt Question 3.5.

Question 3.5

(a) Suppose an eruption occurs from a vent that is 300 m across, with gas and pyroclasts ejected continuously at 300 m s^{-1}. Would you expect this to generate a plinian eruption column or a pyroclastic flow?

(b) The rim of the vent is likely to collapse or be blasted away during the course of an eruption (indeed, this helps to explain the abundance of lithic fragments in pyroclastic deposits). Explain what would happen to the style of eruption if the vent described in part (a) enlarged to one 400 m across (without the exit velocity changing).

(c) Suppose instead that, from the initial conditions in (a), the magma changed to one in which the amount of exsolved water fell to 1 wt %. What would happen in this case?

Column collapse is a common occurrence in the later stages of plinian eruptions, as happened in the archetypal AD 79 plinian eruption of Vesuvius, but it is not inevitable. This is fortunate because the pyroclastic flows generated by column collapse are among the most devastating volcanic hazards.

When material from a collapsing column hits the ground, it transforms into a pyroclastic flow moving at several hundred metres per second that is capable of travelling for 1 to 100 km or more, depending on its initial momentum and amount of material are large enough. When the pyroclastic material is rich in pumice (highly vesicular tephra), then the resulting deposit is known as an **ignimbrite**. An ignimbrite consists of ash interspersed with larger clasts of juvenile pumice and lithic fragments of pre-existing rock ripped off the walls of the conduit. Usually, the flow process results in the deposition of up to three distinct layers, summarised in the generalised cross-section through a complete

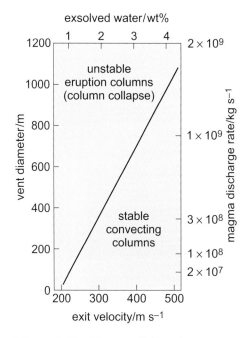

Figure 3.22   The conditions that permit an eruption to develop a convecting plinian eruption column or for the eruption to generate a collapsing column are mapped out in this diagram, according to a mathematical model of eruptions. Note that exit velocity is linked to the percentage of exsolved water, and magma discharge rate is linked to the diameter of the vent.

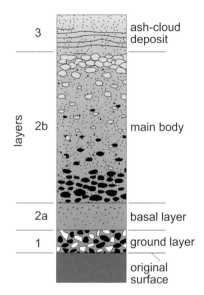

Figure 3.23   Idealised cross-section through an ignimbrite, showing subdivision into layers. Pumice fragments are shown as white shapes and lithic fragments as black shapes. Total thickness may be anything from less than a metre to hundreds of metres.

ignimbrite in Figure 3.23. The lowest layer (known as layer 1 or the ground layer) is deposited from the advancing head of the flow (Figure 3.21) and can be fine-grained or coarse-grained, depending on the dynamics of the flow front region. Above this is the main deposit from the body and tail of the pyroclastic flow (layer 2), and this comprises a thin basal layer (layer 2a) and a thicker upper layer (layer 2b) in which pumice fragments become larger and more concentrated towards the top and lithic fragments become larger and more concentrated towards the bottom.

■   Why do the pumice and lithic fragments show this contrasting distribution pattern?

☐   Pumice is full of vesicles, whereas lithic fragments are dense blocks of rock. Pumice fragments can segregate upwards because of their small density, and lithic fragments segregate downwards because of their large density.

The top-most layer (layer 3 or ash-cloud deposit) is a fine ash deposit, representing material that settled out from the ash column that rose from the flow itself.

Often, the ash and coarser pumice fragments in an ignimbrite are still hot when they come to rest, and may become compacted and welded together under the weight of the upper part of the deposit. The hot pumice fragments are squashed while they are still pliable into forms known as fiamme (pronounced 'fee-am-eh'), which is Italian for flames (Figure 3.24), whereas the colder denser lithic fragments are unaffected.

Figure 3.24   Squashed fragments of pumice, or fiamme, in a welded ignimbrite. The largest in this example is about 15 cm long.

Aside from ignimbrites, the other common kind of pyroclastic flow deposit is a **block-and-ash flow**, which forms from a pyroclastic flow in which the larger clasts are dense blocks (usually of andesite or more felsic composition) rather than pumice. A block-and-ash flow can be triggered by non-explosive collapse from the face of a lava dome, as in the case of the flow shown in Figure 6.9b of Book 1, or be caused by an explosive event that flings out sufficient quantities of blocks (perhaps during a plinian eruption). The deposits from a block-and-ash flow can be distinguished from an ignimbrite chiefly by the lack of pumice.

Sometimes the interior of the collapsed dome is so hot that the moving pyroclastic flow is incandescent and is described as a nuée ardente (pronounced 'noo-eh ardont'). The name, meaning 'glowing cloud' in French, was devised after incandescent pyroclastic flows were observed during the eruption of Mount Pelée on the French Caribbean island of Martinique in 1902 (Figure 3.25).

Question 3.6

Geological field studies show that ignimbrites and other types of pyroclastic flow tend to be thickest in valleys. From what you have learned about pyroclastic flows, how do you explain this observation? How does it contrast with the distribution of pyroclastic fall deposits?

3.4.3 Vulcanian eruptions

The eruptions described in the last two subsections entailed the continuous discharge of magma. But some eruptions occur as discrete explosions – a single or very short-lived explosive release of energy. In these cases, a sudden outburst releases a rapidly expanding cloud of pyroclastic material and gas, which rises into the atmosphere while large blocks are thrown outwards for up to a few kilometres beyond the vent. These ballistically ejected blocks can form small impact craters when they land. Simultaneously, the hot pyroclastic cloud rises to heights of a few kilometres to more than 20 km. This style of eruption is known as **vulcanian**, after the Mediterranean island volcano of Vulcano, whose most recent large eruptions (in the 1880s) exemplify this style of activity. More recent examples have occurred frequently at the Soufrière Hills volcano, Montserrat, during the growth of an andesitic lava dome and they illustrate the general aspects of this eruption style (Figure 3.26).

Vulcanian explosions are due to the buildup of pressure beneath the vent of a volcano that has become blocked by highly viscous lava or debris. Once sufficient pressure has accumulated, the viscous plug explodes, with the result that the underlying magma will suddenly decompress, vesiculate and explosively expand, evacuating the conduit of magma. The conduit can be refilled slowly with viscous magma from depth and may feed a lava dome or again become blocked, possibly leading to another vulcanian cycle.

In contrast to the situation seen at the Soufrière Hills volcano, some vulcanian eruptions are triggered not by magmatic gas but by magma suddenly encountering water. This kind of explosive eruption is driven by the sudden conversion of water into steam and is described as **phreatomagmatic**. This can simply be caused by the magma meeting with groundwater, or by an eruption occurring below less than a few hundred metres of water, or when lava enters a sea or lake.

Figure 3.25   St Pierre, Martinique, a week after the nuée ardente eruption from Mount Pelée in May 1902. The eruption killed 29 000 people and the town was completely destroyed, although only a few centimetres of ash were deposited from the pyroclastic flows that sped through the town. The summit of Mount Pelée is in cloud in the background.

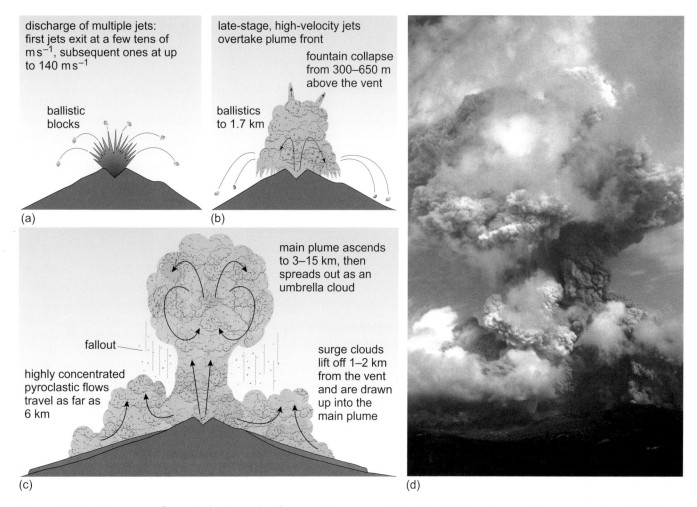

Figure 3.26   Sequence of events during vulcanian eruptions seen at Soufrière Hills volcano: (a) ten seconds (10 s) after the onset of the explosion, multiple jets of ash and ballistic blocks are being ejected; (b) 20 s after explosion onset, a buoyant plume starts to rise; (c) 100 s after onset, pyroclastic flows develop from collapsing parts of the plume, while the buoyant part of the plume ascends several km. (d) Photograph of a developing plume, with pyroclastic flows being fed from the collapsing plume.

3.5   Lahars

Pyroclastic deposits are rapidly emplaced on and around volcanoes, and can be many metres in thickness. Depending on the volcano's environmental setting and frequency of eruption, these deposits may be eroded away or be preserved under later sediments or volcanic deposits. The most catastrophic form of erosion that can happen to recently formed pyroclastic deposits is that torrential rain saturates and then fluidises the deposits, creating a fast-flowing dense suspension not unlike liquid concrete (Figure 3.27). Such flows are known by the Indonesian word **lahar**. They cause devastating destruction to the areas they inundate, and deposit a layer of mud and debris (up to very large boulders in size). Lahars are frequent where explosive volcanism occurs at steep-sided volcanoes in regions with very wet climates – Indonesia is a prime example, as is the Philippines, where about a third of the pyroclastic flow deposits produced by the 1991

(a)

(b)

Figure 3.27 (a) A lahar moving at 3–5 m s^{-1} down a river valley at Pinatubo, Philippines, October 1991. (b) Aftermath of a lahar, Armero, Colombia, November 1985.

eruption of Pinatubo were removed and redeposited by lahars over the years following the eruption.

To produce a lahar, a large amount of water must be able to mix with a large amount of unconsolidated debris at a location where slopes are steep enough to allow speedy flow. So, lahars can be triggered when hot volcanic flows melt and mix with snow or ice. This was the mechanism that generated a lahar at Nevado del Ruiz volcano (Colombia) in 1985, which killed 23 000 people when it engulfed the city of Armero situated 72 km downstream from the volcano (Figure 3.27b).

Volcanic eruptions provide scientists with opportunities to study geological processes at first hand, and you can now apply some of what you have been studying by doing Activity 3.1. Volcanic eruptions are extremely hazardous, so scientists also need to try to predict when eruptions might happen; you can read about this in Box 3.2.

Activity 3.1   Volcanic eruption styles

This video-based activity on the DVD illustrates the processes that occur during effusive and explosive eruptions.

Box 3.2 Forecasting volcanic eruptions

Because volcanoes are dangerous, a goal of volcanology is to forecast when they will erupt and, if possible, what type of activity will occur. But geological studies show that volcanoes are complex, mechanically heterogeneous landforms underlain by an unseen 'plumbing system' in which magma is lodged at imprecisely known depths. Not surprisingly, successfully predicting the time at which magma will erupt at the surface in an eruption is rarely achieved. Indeed, some theoretical models indicate that, as with the weather and financial markets, volcanic systems may, at least some of the time, be inherently unpredictable. None the less, volcanic eruptions typically result from the buildup of pressure within magma below the ground. This can cause the ground to swell upwards and the number of shallow earthquakes to increase as the pressurised system builds up to a state that may cause an eruption. These and other signs of subterranean unrest (such as a change in the chemical composition of gas emitted from the volcano) can be measured using instruments located on the ground (e.g. seismometers to record earthquake activity) or on Earth-orbiting satellites (e.g. infrared thermal detectors to spot anomalously hot areas on volcanoes, and radar to detect small changes in the shape of the ground). The aim is then to use as many of those measurements as possible to understand what is causing the subterranean unrest and whether an eruption might be imminent.

Eruption forecasting is a rapidly developing area of volcanology that has benefited from knowledge gained at volcanoes such as Kilauea (Hawaii), Mt St Helens (Washington State), Mt Pinatubo (Philippines), Soufrière Hills (Montserrat), Etna (Sicily) and Unzen (Japan). One particularly promising method for eruption forecasting is based on the tendency for the number and size of earthquakes at a volcano to increase rapidly in the lead-up to an eruption, as each earthquake reflects a small breakage within the volcano. The size of the earthquake activity can be measured by a quantity called the 'real-time seismic amplitude' (RSAM), which encapsulates in a single number the scale of seismicity averaged over a particular time of a few minutes or hours. It can be calculated continuously – in 'real time' – and followed over time. The RSAM value has been found to increase, becoming extremely large (reflecting a large rate of seismic energy release), in the final moments before an eruption starts. Therefore, a plot of inverse RSAM (i.e. 1 divided by the RSAM value) against time shows a decrease until a value of zero (or close to zero) is reached at the point of failure. The example in Figure 3.28 shows this pattern developing over the four days leading up to a lava dome eruption at Mt St Helens. Starting on 24 May 1985, the inverse RSAM value gradually decreases, allowing the trend on the graph to be extrapolated in order to forecast the time at which the inverse RSAM reaches zero. In this case, the method does a good job of anticipating the time of the eruption that started on 28 May. This approach, developed by Professor Barry Voight of Penn State University, USA, is quite successful in many cases, but does not work every time. There is still much to learn.

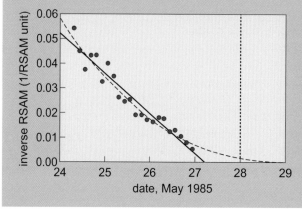

Figure 3.28   A plot of inverse RSAM against time from Mt St Helens. Vertical dotted line is the approximate time at which a lava dome eruption started. Solid and dashed lines are best-fit lines through all of the data.

3.6   Volcanoes

Each of the volcanoes mentioned in the previous sections is the result of many eruptions occurring over their lifetime of many tens or even hundreds of thousands of years. This section considers how the products of individual volcanic eruptions combine to build volcanoes. A volcano's shape depends primarily on the types and relative amounts of effusive and explosive activity that went into its construction, and the amount of erosion that has taken place. In the description that follows, we will refer to basaltic (meaning mafic), intermediate and felsic volcanoes. This is a convenient simplification, indicating the type of magma most commonly erupted, but you should be aware that the average composition may change during a volcano's lifetime. The silica content of magma can even vary during a single eruption.

3.6.1   Shield volcanoes

■   You have read that basaltic volcanism is mostly effusive. Bearing in mind the viscosity of basalt lavas, would you expect basaltic volcanoes to have steep or gentle slopes?

☐   Basaltic lava has very low viscosity (Section 3.1), and so it forms thin flows capable of spreading a long way from their vent. Basaltic volcanoes should therefore be expected to have gentle slopes.

This is indeed the case, and the basaltic volcanoes forming Hawaii's Big Island (Figure 3.29) provide wonderful examples. Their gentle convex profile is reminiscent of a warrior's shield laid face up on the ground, so this shape of volcano is described as a **shield volcano**.

The Big Island of Hawaii is itself built from five overlapping shield volcanoes, and a sixth, Loihi, lies below sea level to the south. Magma erupts from vents at the summit and from vents located along major lines of weakness known as **rift zones** that can be traced downslope from the summit of each volcano (Figure 3.29b). On Mauna Loa for example, many of the lava flows have been fed from vents located along volcanic fissures within the rift zones that run northeast and southwest from the summit, with the result that the shield of Mauna Loa is elongated in this direction. The smaller (and younger) Kilauea volcano is even more elongated.

(a)

(b)

Figure 3.29   (a) Mauna Loa, the largest shield volcano on Earth, rises 8 km above its base and 4 km above sea level and has a diameter of nearly 200 km. Dark lava flows descend from craters along the skyline. (b) Map of the Big Island of Hawaii showing the individual shield volcanoes and their rift zones.

Its southwest and east rift zones can be explained by a tendency for the volcanic edifice to spread towards its unsupported southeast side. Magma ascending beneath the volcano's summit can therefore find easy escape routes along one or other of the rift zones. When an eruption occurs on the volcano's flanks, it may begin by magma breaking out along several kilometres of the rift zone and producing a fine display of fire fountains, but the eruption soon becomes focused at a single point, which becomes the source of most of the lava flows. Over time, the whole volcano is built up from overlapping flow fields emplaced from different rift zones and summit vents.

3.6.2   Stratovolcanoes

When magma of generally intermediate composition erupts to form a volcano, a strikingly different shape results, thanks to a combination of two factors.

1   Intermediate lavas are more viscous than mafic lavas, so they flow less far from their source.

2   Explosive eruptions are more common at intermediate volcanoes than at basaltic ones.

The resulting mixture of short stubby lava flows and pyroclastic flow and fall deposits builds a volcano that is a relatively steep-sided cone, like the classic Mount Fuji in Japan or Mt Shasta, California, which dominates the horizon in Figure 3.7b. The alternation of pyroclastic and lava layers (Figure 3.30) has led to the name **stratovolcano** being applied to this type of volcano. Stratovolcanoes are commonly 10 to 40 km in diameter and 1 to 4 km high. Most volcanoes, including stratovolcanoes, have a roughly circular depression at the summit containing one or more vents. By convention, this is described as a **crater** if less than 1 km in diameter and as a **caldera** if bigger than this. A crater or caldera is usually considerably wider than the vent and the conduit that feeds it, and is formed by a mixture of collapse and subsidence on faults. Stratovolcanoes have summit craters or calderas up to a few kilometres in diameter.

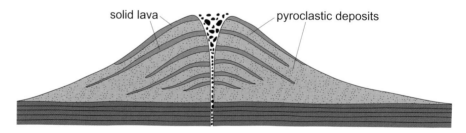

Figure 3.30   Diagrammatic cross-section through a stratovolcano (not to scale).

3.6.3   Collapsing volcanoes

Any mountain is inherently unstable and, unless it is strong and internally coherent, gravity will cause the weakest part to collapse or spread sideways. A volcano is particularly vulnerable because its structure can be weakened by the repeated injection of magma and the rotting away of the rocks inside the volcano as a result of hydrothermal activity. Not every volcano is capable of collapsing, but when they do, material on the steep upper slopes avalanches onto the flanks of the volcano and beyond (Figure 3.31). Collapse may be triggered by the arrival of magma beneath a vulnerable part of the volcano, as happened under Mount St Helens (USA) in May 1980. There, the northern sector of the volcano disintegrated, collapsed and avalanched a total distance of 25 km, with the immediate consequence that magma within the volcano was suddenly 'uncorked'. The exploding magma sustained a 16 km-high plinian eruption column for 9 hours. In other situations, collapse can be triggered by earthquakes, with no magma being involved at any stage. The **debris avalanche** deposits formed in these ways contain very large blocks (up to km scale) of the original volcanic edifice embedded in a mixture of finer broken fragments and have a characteristically hummocky surface (Figure 3.31).

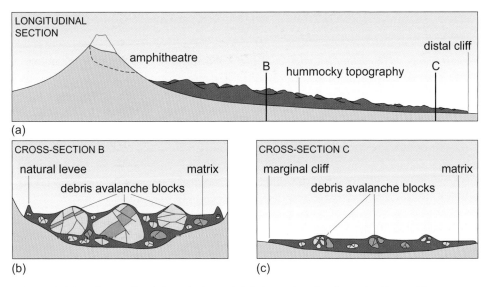

Figure 3.31   Generalised description of debris avalanche deposits formed by collapse of a steep volcano: (a) incomplete collapse of the volcano's summit leaves an amphitheatre and generates an extensive avalanche deposit (dark brown); (b) levees near the edge of the deposit and large blocks, whose size decreases downstream (c).

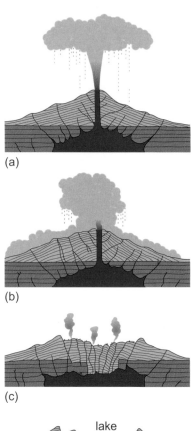

(a)

(b)

(c)

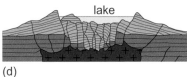

lake

(d)

Figure 3.32   The cataclysmic eruption of about 55 km³ of felsic magma about 7700 years ago produced Crater Lake caldera, Oregon, and destroyed an ancestral andesitic stratovolcano. The caldera-forming eruption started with (a) a plinian eruption that produced a widespread fall deposit, followed by (b) emplacement of ignimbrites from pyroclastic flows. (c) Partial evacuation of the magma chamber caused its roof to collapse, forming an 8 km-diameter caldera. (d) The situation today. There have been some small andesite eruptions since the caldera was formed. The caldera is now occupied by a lake of water, and the remains of the magma chamber have probably solidified.

3.6.4   Ignimbrite calderas and supervolcanoes

Where felsic magma is erupted in abundance, the resulting landform is quite unlike a picturebook stratovolcano. In the rare instances where eruption of felsic magma is effusive, the result is thick lava flows like those shown in Figure 3.7. However, as you have seen, the high viscosity and high water content of felsic magmas means that most felsic eruptions are explosive. The largest volume felsic eruptions produce ignimbrites. Eruptions of at least 1 to 10 km³ of magma usually evacuate a shallow magma chamber sufficiently for the roof of the chamber to collapse into the chamber, forming a caldera. This is what happened during the formation of Crater Lake (USA) some 7700 years ago (Figure 3.32) and more recently, on slightly smaller scales, during the eruptions of Tambora (Sumbawa, Indonesia) in 1815, Krakatau (west of Java, also in Indonesia) in 1883, and Pinatubo (Philippines) in 1991.

However, these devastating examples were much smaller than the largest caldera-forming events that have occurred in the geological past. For example, eruption of 5000 km³ of felsic magma about 28 million years ago from a volcanic centre in Colorado (USA) generated an extensive ignimbrite known as the Fish Canyon Tuff (tuff is an old geological term for pyroclastic rock) and produced a caldera 35 km across and 75 km long. More recently (74 000 years ago, abbreviated as 74 ka) a similar type of event generated the Youngest Toba Tuff and the Toba caldera when 2800 km³ of felsic magma erupted in northern Sumatra (Figure 3.33).

In extremely large examples such as these, ignimbrite with a thickness of about 100 m smothers an area of several thousand square kilometres. The ignimbrite that accumulates in the foundering caldera can be up to 2 km thick; a rough rule of thumb is that one-third of the magma is deposited outside the caldera and two-thirds inside the caldera. Long after an ignimbrite-forming eruption, lava flows or domes are usually erupted intermittently within the caldera over a period that may last several hundred thousand years. If new magma is injected into the magma chamber, the floor of the caldera will be domed upwards to form a broad rise within the caldera that is described as a resurgent centre.

Very large eruptions are clearly at the extreme end of the spectrum of volcanic eruptions, in terms of size and effects. They have been called 'supereruptions' and, in 2000, a BBC television documentary programme coined the term 'supervolcano'. Since then a **supereruption** has been defined as an eruption producing more than 10^{15} kg of magma (in terms of volume, this is 450 km³). A **supervolcano** can be defined as any volcano that has produced at least one supereruption. Large volume eruptions are rare, and supereruptions occur *on average* about once every 100 000 to 200 000 years. The most recent supereruption occurred in New Zealand about 26 000 years ago and formed a caldera now occupied by Lake Taupo.

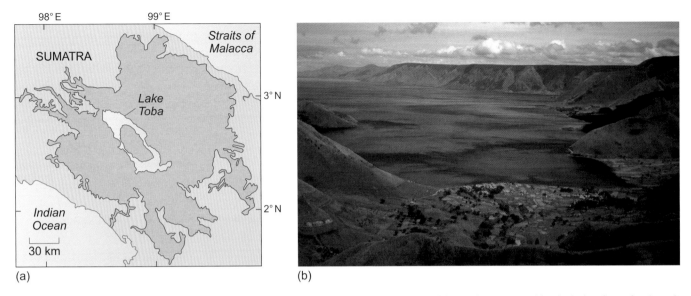

Figure 3.33 (a) Map of northern Sumatra showing the outline of the Toba caldera, the extent of ignimbrite deposits (grey) produced in this eruption and the resurgent centre that forms an island at the centre of the caldera lake. (b) The view from the northwest corner of Lake Toba, showing cliffs of welded ignimbrite forming the caldera walls.

3.7 Summary of Chapter 3

1 Magma is liquid rock, often containing some crystals and, especially at low pressures, some gas bubbles. For a given magma composition, increasing temperature and increasing volatile content will cause the viscosity to decrease. Increasing silica content and crystal content (at constant temperature and volatile content) cause viscosity to increase. Generally speaking, composition, temperature and volatile content are correlated such that the volatile content of a magma goes up with increasing silica content, as does viscosity.

2 The surface textures of lava flows depend on viscosity, rate of flow and rate of cooling. These factors also affect the shape of a lava flow field.

3 Explosive eruptions occur when decreasing pressure allows volatiles to exsolve and form bubbles. The more viscous the magma, the harder it is for bubbles to escape, so expansion of bubbles in viscous felsic magma can create a particularly violent explosion. Easier escape of bubbles from low-viscosity basaltic magma favours strombolian explosions and fire fountains.

4 Pyroclastic fall deposits are produced by fallout of tephra from explosive eruptions that generate high plumes (plinian eruptions) and they blanket topography. The thickness of the deposit and the size of the clasts both decrease away from the volcano in patterns that depend on the wind direction and discharge rate.

5 Pyroclastic flows can originate by collapse of an eruption column or by an avalanche from an unstable lava dome. They flow downhill at high speed and tend to fill valleys. Pyroclastic flows deposit ignimbrites and block-and-ash deposits.

6   The shape of a volcano depends on the dominant kind of magma erupted. Mafic eruptions produce shield volcanoes (dominantly basaltic lavas), intermediate eruptions produce stratovolcanoes (interlayered andesitic lava and ash), and felsic eruptions of more than 1 to 10 km³ are commonly caldera-forming. Calderas are formed by subsidence of the roof of a magma chamber when it is partly emptied during a large-volume (usually explosive) eruption, or a series of smaller eruptions.

7   Volcanoes are weak edifices, and may collapse to form debris avalanche deposits.

3.8   Objectives for Chapter 3

Now you have completed this chapter, you should be able to:

3.1 Describe and recognise the forms taken by flowing lava on dry land and under water, and the factors controlling these.

3.2 Explain the links between magma type, gas content and the style of eruptions, and describe and recognise the main processes occurring during different styles of eruption, giving examples of the resulting deposits.

3.3 Recognise examples of different kinds of volcanoes and volcanic features, and explain how their shapes are controlled.

Now try the following questions to test your understanding of Chapter 3.

Question 3.7

Indicate the eruption products or eruption style you would expect for mafic and felsic magma of high and low volatile content as indicated in Table 3.2 by matching each case with one of (a)–(d) from the list:

(a) lava domes   (c) pahoehoe and a'a lava flows

(b) fire fountains   (d) ignimbrites and pyroclastic fall deposits.

Table 3.2   For use with Question 3.7.

	Low volatile content	High volatile content
mafic magma		
felsic magma		

Question 3.8

Discuss how, on the basis of a single outcrop in the field, you could identify a deposit as an ignimbrite.

Chapter 4   Intrusive igneous rocks

This chapter considers igneous bodies produced by intrusion of magmas that did not reach the surface. It begins by looking at the pathways taken by magma in the shallow sub-volcanic level, and then investigates deeper, larger bodies of intrusive rock that represent solidified magma chambers.

4.1   Minor intrusions

Magma that erupts onto the Earth's surface and produces extrusive igneous rocks clearly comes from below the ground, but what form does the magma transport system take, and can this system be recognised in the geological record, on maps and in the field? By their very nature, it is impossible to observe directly the subsurface magma conduits supplying an ongoing volcanic eruption. So, as with many aspects of geology, various bits of indirect evidence have to be pieced together in order to solve the problem.

- What approaches could be used to gather evidence about the shallow roots of volcanoes?

- There are three ways of doing this: interpreting observations of active vents made during or after an eruption; interpreting geophysical signals such as earthquakes and surface deformation that occur during an eruption; and interpreting rocks exposed by erosion of old volcanoes.

In some eruptions, such as those along the rift zones of Hawaii's volcanoes and in Iceland, activity occurs along a row of craters that can be many kilometres long (Figure 4.1a). The simplest explanation for this is that directly beneath the eruptive fissure must lie a long narrow vertical intrusion feeding magma upwards to the active vents.

- What name is given to an intrusion with this shape?

- A dyke.

Recall that a dyke is a discordant curtain-like intrusion (Figure 4.2), intruded in a near-vertical plane, whereas a sill is a sheet-like body, originally intruded horizontally, that is mostly concordant with the bedding of the strata it intrudes.

Many of the eruptions in Iceland involve fire fountaining from vents located along fissures. Others erupt from vents at centres of activity associated with hydrothermal activity and calderas located at the centre of fissure swarms (Figure 4.1b). The central volcanoes are thought to overlie persistently hot magma chambers. Each central volcano and its associated fissure swarm define a separate volcanic system.

One of northern Iceland's volcanic systems, Krafla, was active between 1975 and 1984, with much seismic activity, ground deformation and fire fountaining. The activity was cyclic, with slow inflation of the ground above the central volcano followed by fast deflation and migration of earthquakes away from the centre of ground deformation. This pattern was repeated 21 times, with fissure eruptions happening 9 times during periods of deflation and earthquake migration.

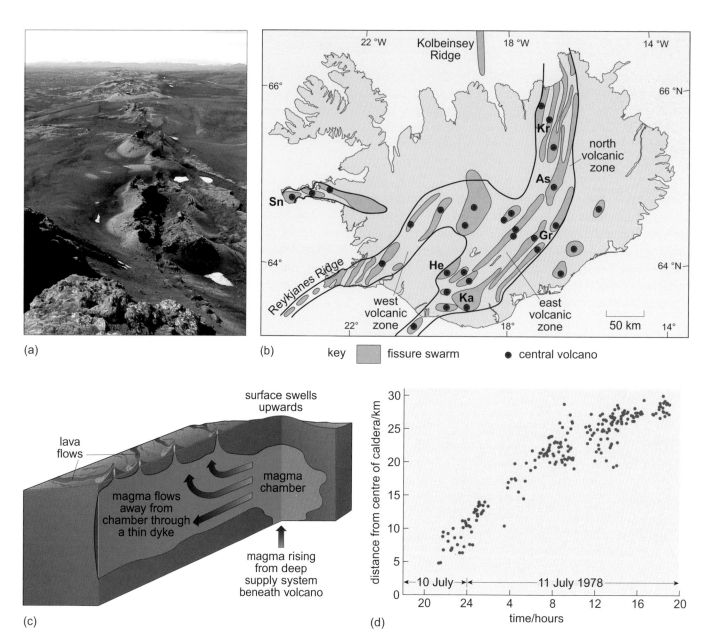

Figure 4.1   (a) Aerial view of the 27 km-long Laki crater row, southern Iceland, which erupted basaltic scoria and lava in 1783–1784 in one of recorded history's largest volcanic eruptions (involving about 15 km³ of magma). (b) Map of Iceland showing locations of central volcanoes and fissure swarms arranged in volcanic zones across the country. Selected volcanic systems identified by abbreviations: As, Askja; Gr, Grimsvotn (source of the Laki eruption); He, Hekla; Ka, Katla; Kr, Krafla; Sn, Snaefellsjokull. (c) Interpretive model of vertical magma rise beneath a volcano and lateral emplacement of a dyke, followed by eruption along a crater row. (d) Plot of earthquake location, measured northwards from the centre of the Krafla caldera during 10–11 July 1978.

Figure 4.2   Erosion has exposed the three-dimensional view of a dyke, which forms the jagged crest of the ridge emanating from a larger igneous intrusion known as Ship Rock (New Mexico, USA). The dyke varies from 0.5 to 5 m in width. A second dyke is visible in the background, trending to the left of the photograph.

The activity at Krafla can be explained if the volcanic system comprises a shallow magma chamber into which more magma is injected from below (Figure 4.1c). This causes the chamber to expand and the ground to swell upwards until the strength of the enclosing rocks is exceeded. At this point, fractures open up in the surrounding rock, allowing magma to escape from the chamber into a newly forming dyke. The ground surface deflates and the dyke extends in length by propagating away from the chamber, causing small earthquakes wherever the crust is broken apart at the tip of the advancing dyke (Figure 4.1d). In some cases magma breaks through to the surface at points above the dyke, giving rise to fire fountains.

Question 4.1

For the case of the dyke emplacement event at Krafla (Figure 4.1d), estimate the approximate speed at which the dyke advanced. Was the speed constant?

Dykes rarely occur in isolation, and many areas of the crust have been repeatedly intruded by dykes, by magma intruding either from below or laterally from a shallow chamber. For example, deeply eroded areas of eastern Iceland expose parallel dykes up to a few metres wide cutting vertically through basaltic lava flows and making up to 8% of the crust. These appear to be the subsurface equivalent of fissure swarms that characterise present-day volcanic systems such as Krafla. The parallel orientation of the dykes indicates that the path of least resistance taken by the intruding magma was always in the same direction. A reasonable explanation for this is that the crust was consistently being pulled apart in the direction perpendicular to the strike of the dykes. The large array of dykes is known as a **dyke swarm**.

In the absence of external influences, dykes will be radially distributed about a central source, as found for example in the eroded cores of stratovolcanoes in regions where the crust has not been subject to tensional forces in a preferred direction (Figure 4.3a). Radial dykes may reach the surface of the flanks of an active volcano (in which case they may act as sources for eruptions), but are more clearly seen after a long period of deep erosion.

Dykes are important means by which magma moves through the crust and to the surface. But other types of intrusion can be found, especially horizontal intrusions, which are known as sills. Sills may be metres to hundreds of metres

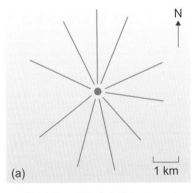

(a)   1 km

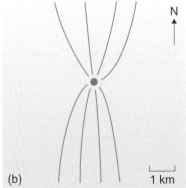

(b)   1 km

Figure 4.3   Schematic maps of the outcrop patterns of: (a) radial dykes around a volcano intruded in the absence of regional stress; (b) a dyke swarm associated with a central volcano in a region of east–west extension.

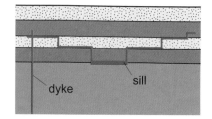

Figure 4.4   Cross-section showing the typical relationship between a sill (here fed by a dyke on the left) and horizontal strata intruded by it. The sill is generally concordant, but is locally discordant where it cuts up or down between bedding planes.

thick, and usually occur only at very shallow levels within the crust where mafic magma has been injected into horizontal or gently dipping sedimentary strata within a kilometre or so of the surface (Figure 4.4).

If dykes intrude as vertical 'walls' of magma because the surrounding rock can be most easily pushed (or pulled) aside horizontally, then sills intrude because it is energetically easier to lift the overburden of rocks than to shoulder aside the country rock.

When dykes and sills intrude, the enclosing country rock becomes heated by the magma, but the total amount of heat escaping from a minor intrusion is generally insufficient to cause extensive contact metamorphism of the rocks it invades. Thus, sedimentary rock may be baked and become hard and brittle in a zone a metre or so wide against the contact with a dyke or sill, but growth of new, metamorphic minerals is uncommon.

In contrast, the consequences of cooling within the intrusion can be more noticeable. Intrusions lose heat to the country rock and therefore solidify from the outside in. As the magma cools and solidifies into igneous rock it shrinks, with the result that contraction cracks develop, penetrating into the intrusion to generate hexagonal columns arranged parallel with the cooling direction (and perpendicular to the cooling surface). This is why dykes and sills often have regular six-sided columnar joints arranged perpendicular to their margins (Figure 4.5). For the same reasons, columnar joints can be found in some lava flows.

Figure 4.5   Vertical columnar joints in a 30 m-thick dolerite sill at Kilt Rock, Skye, formed by thermal contraction as the sill cooled through its roof and floor. The contact between the base of the sill and underlying horizontal sandstone strata is roughly halfway down the cliff. A second sill is also exposed near the base of the cliff, but columnar jointing is much less well developed. Reproduced with the permission of the British Geological Survey © NERC. All rights Reserved.

The loss of heat from a dyke or sill that has been emplaced into cold country rock is sufficiently rapid to affect the texture of the igneous rock within the minor intrusion. The magma at the margin is cooled so rapidly that only a few small crystals have time to form, and the texture of the igneous rock at the contact is often glassy. Minor intrusions are therefore characterised by **chilled margins** and interiors with medium grain size. A thin section through the chilled margin at the base of the Whin Sill in NE England is shown in Figure 4.6. The field of view is too small to show any of the medium-grained, completely crystalline interior of the sill.

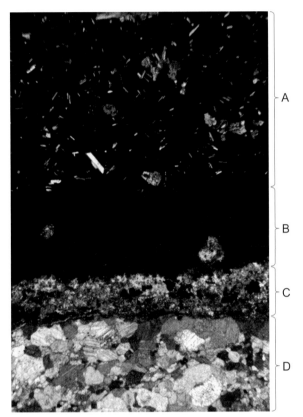

A

B

Figure 4.6   Thin section through the base of the Whin Sill (viewed between crossed polars), showing an area about 1.4 mm long. In region A, crystals of pyroxene and plagioclase feldspar occur in a fine-grained and partly glassy matrix. Region B is the actual chilled margin, which is glassy throughout except for a couple of olivine crystals that presumably grew within the magma before it was emplaced. Regions C and D are within the country rock, which in this case is limestone. You should recognise the pastel interference colours of the calcite crystals in D. In region C (only about 0.2 mm wide), some new metamorphic minerals have begun to grow because of the intense heat from this >50 m-thick sill.

C

D

Question 4.2

Figure 4.7 shows black and white images of thin sections of two samples from the Cleveland Dyke, County Durham, England, collected from where the dyke is about 20 m wide. Look at these, and then attempt the following.

(i)   Decide whether the average grain sizes of (a) and (b) in Figure 4.7 are coarse, medium, fine or glassy.

(ii)   Which of the two thin sections comes from closer to the margin of the dyke?

(iii)   How would you explain the presence of the relatively long (0.3 mm) plagioclase feldspar crystal visible in each specimen?

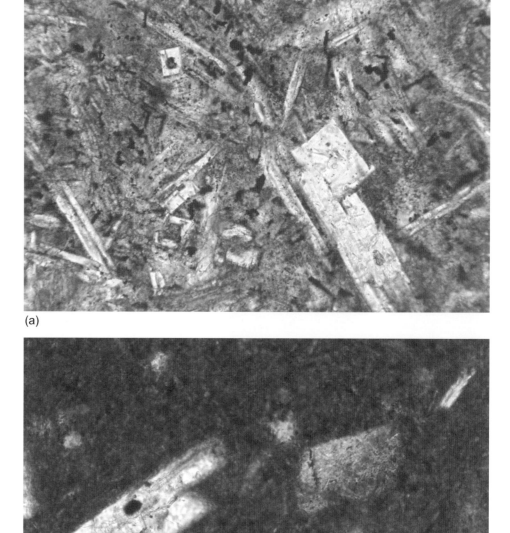

(a)

(b)

Figure 4.7   (a) and (b) Thin sections from two parts of the Cleveland Dyke (plane-polarised light, photographed in black and white) (field of view 0.7 mm across). For use with Question 4.2.

Sills and dykes are minor intrusions that are characteristically planar in shape. In outcrop, dykes typically define straight lines that cut across topography as a result of being vertical, whereas sills follow contours as a result of being flat-lying. There are some minor intrusions, however, that can have other outcrop patterns. Cone sheets and ring dykes are dykes with approximately circular outcrop patterns (so they depart from the definition of a dyke as a planar body). **Cone sheets** converge downwards towards the top of the magma chamber. They are usually narrow features a few metres or less in width. Conversely, a **ring dyke** dips steeply outwards and is typically anything from a hundred metres to a couple of kilometres wide. The geometry of these two kinds of minor intrusion is illustrated in Figure 4.8; study this and then attempt Question 4.3.

Question 4.3

Bearing in mind that emplacement of dykes of any kind requires displacement of the country rock, how do the blocks of the rock labelled A on Figure 4.8 move relative to the magma chamber during emplacement of (a) cone sheets, (b) a ring dyke? In each case, decide whether this implies that the minor intrusion was created by forceful injection or passive intrusion of magma.

Activity 4.1   Minor intrusions on geological maps

In this activity, you will use geological maps to deduce how and when some different types of minor intrusion were emplaced in Britain's geological past.

4.2   Plutons

The minor intrusions discussed in the previous section are typically 0.1 to 100s of metres thick. Larger bodies of intrusive igneous rock, of kilometre-size or more, are known as plutons and are produced by solidification of substantial bodies of magma at depths of several kilometres or even tens of kilometres.

■ Bearing in mind the depth at which a pluton crystallises, would you expect it to be fine, medium or coarse grained?

☐ A pluton crystallises deep below the Earth's surface, therefore it cools slowly and the crystals have time to grow to a large size. Plutons are typically coarse grained.

■ Bearing in mind its grain size, what rock type would you expect to find in a pluton of felsic composition?

☐ Granite, according to the classification scheme you met in Table 6.2 of Book 1.

■ Why are plutonic rocks such as granite now exposed at the surface?

☐ They are exposed as a result of erosion.

Next, we will look at examples of granite plutons and their relationships with the country rock in cases representing different depths within the crust, beginning with 'deep' granite plutons emplaced at mid-crustal levels. At this depth (about 20 km), most continental crust has a felsic or intermediate composition and the

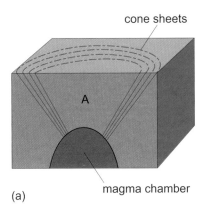

(a)

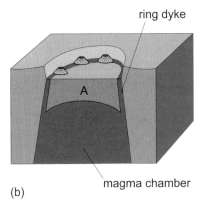

(b)

Figure 4.8 Diagrammatic cross-sections through (a) a series of cone sheets and (b) a ring dyke. The circular depression bounded by the ring dyke in (b) is an example of a caldera.

temperature and pressure are generally sufficient for the rock to be regionally metamorphosed to felsic and intermediate gneisses and schists with well-developed metamorphic foliation. When a gneiss is subjected to conditions at which melting begins, a small amount of partial melting can turn its texture into that of a migmatite (Book 1, Figure 8.5) containing a pervasive network of granite veins. The deepest granite plutons therefore tend to consist of elongate masses of granite aligned parallel to the metamorphic foliation of the country rock (Figure 4.9a), which is a sign that within these hot crustal regions granite magmas can form, collect into larger bodies and start to move around, locally intruding the migmatite. Often the final stage of deformation of the country rock outlasts the intrusion of the granite, so that the granite itself has a foliation imposed upon it.

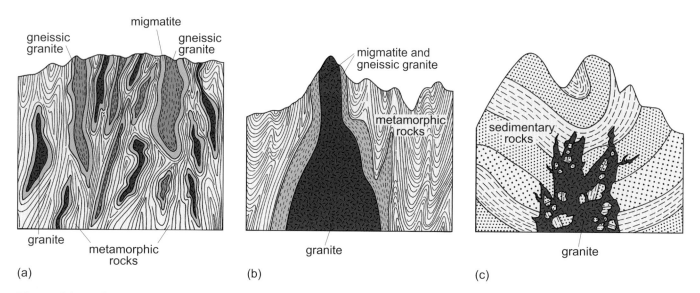

Figure 4.9   Schematic cross-sections of granite intrusions emplaced at different crustal depths and now exposed by erosion: (a) deep (more than about 20 km) mid-crustal intrusions associated with migmatites; (b) a concordant pluton intruded at about 10 to 20 km depth; (c) a discordant pluton intruded at shallow depth (less than about 10 km).

At somewhat shallower depths it is more common to find a single large pluton rather than many small ones (Figure 4.9b), but the plutons still tend to be parallel to the foliation within the country rock, rather than cross-cutting it. In other words, these plutons tend to be concordant to the foliation of metamorphosed country rock, rather than discordant with it. This is because at this depth the country rock was hot enough (and therefore deformable enough) for a pluton-sized mass of granite to force its way upwards by pushing the country rock aside (possibly aided by assimilation of country rock into the magma). A pluton that has risen in this way is described as a **diapir** and might be envisaged as a balloon of magma creeping upwards through the crust until it can move no further.

Plutons believed to have been emplaced at still shallower levels in the crust (<10 km depth) can usually be seen to be discordant. Their edges cut across the fabric of the country rock, which at this depth is usually sedimentary bedding (Figure 4.9c). When intruding into cold, brittle crust, a shallow-level pluton cannot push the country rock aside and force its way up diapirically, as may sometimes be the case deeper down. Nor can it simply melt its way upwards by assimilating all the country rock in its path, because this would require more heat than the pluton contains.

Given that assimilation cannot be a major factor, space must somehow be generated within the crust for a pluton to occupy. As you have seen, the middle crust may be deformable enough to be pushed aside and deform around a diapir. However, plutons in the upper crust can often be seen to occupy spaces made available by fracturing and fault movement where the crust was being pulled apart. It is this pulling apart that often opens up space into which magmas can intrude (Figure 4.10). A further example of fault-controlled intrusion is a **ring fracture** that allows a cylinder of country rock to subside while the magma rises up the fracture and collects in an overlying magma chamber where it solidifies as a pluton (Figure 4.11). In addition, particularly shallow plutons may be accommodated by a certain amount of updoming of the surface.

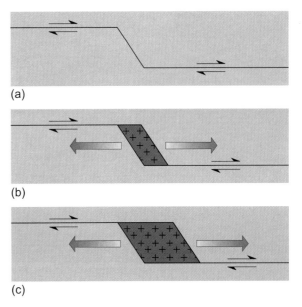

Figure 4.10 Sketch maps showing how movement along parallel faults can open up space that can be occupied by magma intruding from below.

Field evidence suggests one more process by which plutons may sometimes intrude the shallow crust. Figure 4.12a is a photograph of the roof zone of a pale granite pluton that has been intruded at shallow level into darker country rock, resulting in angular blocks of country rock being broken off into the intruding magma. A chunk of country rock enclosed within an intrusion is an example of a **xenolith** (xeno, pronounced 'zeno', is Greek for 'foreign', and the xenoliths are 'foreign rocks' in the sense that they are unrelated to the origin of the igneous rock in which they are found).

The process by which xenoliths are plucked away from the country rock by the magma is known as **stoping** (pronounced 'stow-ping'). Some xenoliths may sink to the bottom of the pluton. Others can sometimes be identified near the centres of plutons, where they tend to have more rounded shapes, and sometimes a recrystallised fabric, indicating that the heat from the surrounding magma was sufficient to cause them to become soft and mushy. In extreme cases, xenoliths are no more than ghostly dark patches (Figure 4.12b), showing that they have become almost entirely assimilated into the granitic magma. It can be difficult to distinguish these types of xenolith from patches of other magmas that have been incompletely mixed with the main granite magma.

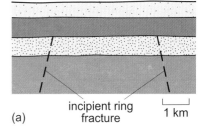

(a) incipient ring fracture   1 km

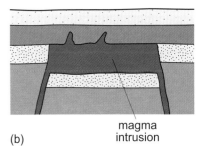

(b) magma intrusion

- ■ You have seen that the heat from a granite can have extreme effects on xenoliths that find their way into the heart of the intrusion. But what effect, if any, would you expect to see in the country rock in contact with the granite?

- □ You would expect heat to have been conducted from the granite into the country rock, and this could have caused contact metamorphism.

The zone of contact metamorphosed rock around a shallow pluton is described as its metamorphic aureole. This is alternatively known as a 'contact aureole' or 'thermal aureole' and can be of the order of a kilometre wide. The heat of

Figure 4.11 Cross-sections to show emplacement of a shallow-level pluton by subsidence of a cylinder of country rock bounded by a ring fracture. (a) A ring fracture forms, which would be approximately circular in plan view, but in this example does not reach the surface. (b) The cylinder of country rock bounded by the ring fracture subsides, and magma rises up to occupy an overlying magma chamber.

(a)   (b)

Figure 4.12   Examples of xenoliths in two granite plutons. (a) Part of a granite pluton, near its upper contact, showing an abundance of dark angular xenoliths broken from the enclosing country rock. (b) A xenolith with indistinct margins in granite. (Vertical marks are drill holes related to quarrying.)

the pluton may also cause water within the crust to convect by hydrothermal circulation, in which hot water is expelled through the roof of the intrusion and colder water is drawn in through its sides. This fluid dissolves and reprecipitates certain elements, some within the pluton and some within the country rock, and the process may continue long after the pluton has solidified. Such movement of elements by hot solutions (sometimes referred to as hydrothermal fluids) can cause further mineralogical changes in both the pluton and the country rock – an effect that is described as **metasomatism**. It may also lead to deposition of ore minerals, particularly in veins within and around the pluton, sometimes in economic quantities.

To describe and understand the geology of plutonic rocks, it becomes necessary to have a classification scheme for the different rock types that are encountered. The one in Figure 4.13 is based on a compositional classification according to silica content as reflected by the abundances of quartz, feldspars and mafic minerals. You met this diagram in Book 1 and it expresses a gradation between felsic and ultramafic compositions. Although silica content is the most important variable in igneous rock composition, it is not the only one. Perhaps the next most important factor (and one which can vary independently of silica content) is the relative abundances of the alkali metals sodium (Na) and potassium (K).

■   What common mineral group contains large amounts of Na and K?

☐   The feldspars, particularly the sodium-rich plagioclase feldspar albite ($NaAlSi_3O_8$) and potassium feldspar ($KAlSi_3O_8$).

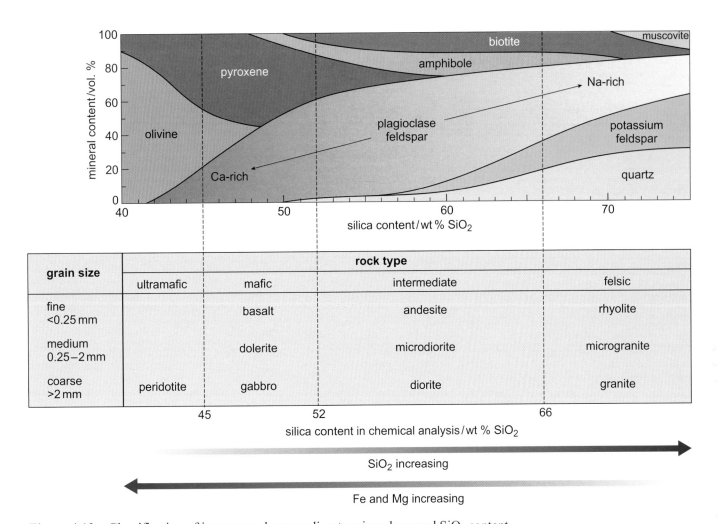

Figure 4.13   Classification of igneous rocks according to mineralogy and SiO$_2$ content.

Plagioclase feldspar exists as a solid-solution series of any composition between the two end-members albite (NaAlSi$_3$O$_8$) and anorthite (CaAl$_2$Si$_2$O$_8$).

■   According to Figure 4.13, which types of coarse-grained igneous rocks have plagioclase feldspar, quartz and alkali feldspar adding up to more than 50% of the rock?

☐   All granites and many diorites have these three minerals in abundances that add up to more than 50%.

The importance of quartz, plagioclase feldspar and alkali feldspar has led geologists to devise the classification system for coarse-grained igneous rocks shown in Figure 4.14. This requires the geologist to be able to distinguish and estimate the relative proportions of the three felsic minerals alkali feldspar, plagioclase feldspar and quartz. This is best done using a petrological (polarising) microscope, as alkali feldspar and plagioclase feldspar are not always easily distinguished in hand specimen. Figure 4.14a is often known as a **QAP diagram**, after the initial letters of the three minerals on which it is based.

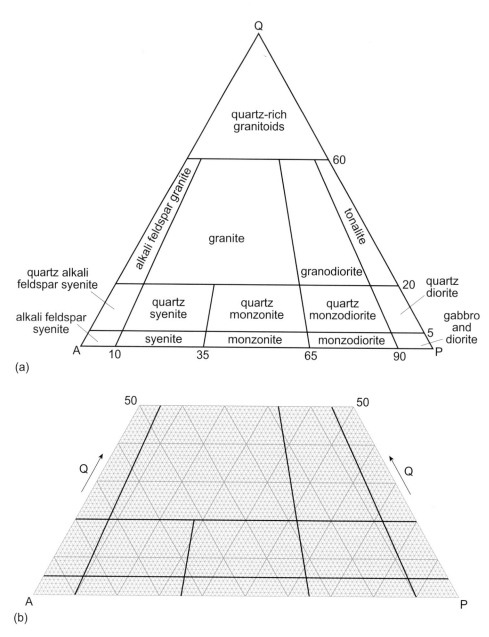

Figure 4.14 (a) The QAP diagram for plutonic igneous rocks according to the relative percentages of quartz (Q), alkali feldspar (A) and plagioclase feldspar (P). Medium-grained rocks take the name of their coarse-grained equivalents with the prefix 'micro' added (e.g. microsyenite), except that a 'microgabbro' is usually called a dolerite. Gabbro is distinguished from diorite by the composition of its plagioclase feldspar, which is >50% albite in diorite and <50% albite in gabbro. (Note: all these rock types have other minerals present, notably mafic minerals, but this classification is based only on the quartz and feldspar content.) (b) The lower half of (a) with the boundaries between the fields of the different rock types superimposed on a regular grid of triangular plotting paper.

■ A plutonic rock contains 17% quartz, 13% alkali feldspar, 42% plagioclase feldspar, 7% hornblende, 16% biotite and 5% accessory minerals. Calculate the relative percentages of quartz, alkali feldspar and plagioclase feldspar. According to Figure 4.14, what is the name of the rock?

□ The percentage quartz value (Q) is:

$$\frac{17}{17+13+42} \times 100\% = 24\% \text{ (to 2 significant figures)}$$

The percentage alkali feldspar value (A) is:

$$\frac{13}{17+13+42} \times 100\% = 18\% \text{ (to 2 significant figures)}$$

The percentage plagioclase value (P) is:

$$\frac{42}{17+13+42} \times 100\% = 58\% \text{ (to 2 significant figures)}$$

The rock is therefore classified as a granodiorite.

To reassure yourself that you can use Figure 4.14 to classify igneous rocks, try Question 4.4.

Question 4.4

What name would you give to igneous rocks containing the relative proportions of quartz, alkali feldspar and plagioclase feldspar given below, bearing in mind the grain size? (Note: the values given here have already been recalculated so as to exclude the proportions of other minerals.)

(a) Coarse-grained: 3% quartz, 72% alkali feldspar, 25% plagioclase feldspar.

(b) Medium-grained: 30% quartz, 15% alkali feldspar, 55% plagioclase feldspar.

(c) Coarse-grained: 25% quartz, 5% alkali feldspar, 70% plagioclase feldspar.

(d) Coarse-grained: 3% quartz, 2% alkali feldspar, 95% plagioclase feldspar (of composition 30% anorthite, 70% albite).

Very few plutons are composed of a single rock type. There is usually a range of compositions that is sufficiently broad to cover two or more of the rock types in Figure 4.14 – and Figure 4.15 shows an example of this.

The Rogart intrusion (Figure 4.15) displays another feature that is characteristic of many granitic plutons, in that it consists of successive units intruded into one another. The first unit to be emplaced in the Rogart intrusion was the one mapped as tonalite. The granodiorite was then intruded into the centre of this mass, with a steeply dipping contact. This in turn was intruded by the unit mapped as granite. It is common for the compositions of successive intrusions to change in this manner, from generally dioritic or tonalitic (intermediate) at first to generally granitic (felsic) towards the end.

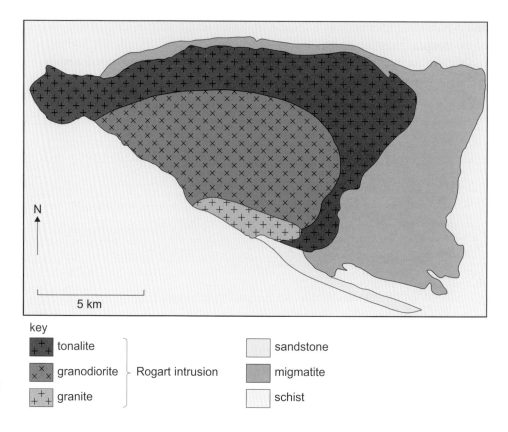

Figure 4.15   A geological map of the Rogart intrusion, Sutherland, Scotland (centred near grid reference NC (29) 6907). Contacts between the different plutonic rock types are gradational over tens to hundreds of metres.

key

tonalite		
granodiorite	Rogart intrusion	
granite		

sandstone

migmatite

schist

The final stages of crystallisation of a plutonic rock are controlled by the fact that magma can hold a greater percentage of volatiles than can solid rock. As crystallisation proceeds, the remaining melt becomes progressively richer in volatiles. The volatile elements (or any other elements) that are not incorporated into the growing crystals are said to be incompatible elements. The last fraction of a per cent to crystallise may be exceptionally rich in volatiles and other incompatible elements and form veins of coarse-grained pegmatite (Book 1, Figure 6.13a). Alternatively, sudden escape of the volatiles can trigger rapid crystallisation of a fine- to medium-grained sugary-looking mixture of quartz and alkali feldspar referred to as aplite (Book 1, Figure 6.13b).

Our account of plutons has so far concentrated on granites and related intermediate rock types. Mafic magmas also form plutons. In the UK, intrusions of gabbro are found in western Scotland, notably on the islands of Rum, Mull and Skye, and were emplaced about 60 million years ago.

■   Which minerals are common in gabbro?

☐   Plagioclase feldspar, pyroxene and olivine.

In some cases the rock shows a striking horizontal layering, defined by different proportions of minerals. World-class examples of this are found on Rum, where vertical variation in the proportions of these minerals (and in the composition of minerals within their solid-solution series) takes place on a scale of metres to tens of metres (Figure 4.16); it is an example of a **layered intrusion**. The cause of the layering is not always clear, but in some cases crystals have accumulated from a body of cooling basaltic magma, with early-formed crystals sinking to the floor of the magma chamber similar to sedimentary grains settling through water. In other cases, new drafts of magma have entered the chamber and, being denser than the resident magma, have spread over the chamber floor, solidifying into a layer of rock with a different composition (and different mineral proportions) to the original magma.

(a)   (b)

Figure 4.16 (a) View of Hallival, Rum, Scotland showing prominent horizontal layering of mafic and ultramafic rock. The resistant layers are gabbro, whereas the more easily weathered layers are richer in olivine. Note the rocks forming the low-lying area behind the mountain are also layered, and dip to the left (towards the northwest), but these are the sedimentary country rocks into which the gabbro was intruded. Reproduced with the permission of the British Geological Survey © NERC. All rights Reserved. (b) Well-developed mineral layering in peridotite, sandwiched between gabbro layers, Rum, Scotland. Peridotite layer is 2.5 m thick.

4.3 Batholiths

Most geologists would regard the Rogart intrusion as a single pluton, even though it is a concentric complex constructed by the arrival of successive pulses of magma. Likewise, layered intrusions with evidence of several separate batches of magma being intruded over time are still considered to be plutons. Sometimes two or more neighbouring plutons overlap (the younger intruding the older pluton) in a more complicated manner than the concentric pattern exemplified by the Rogart intrusion. Such a combination of plutons is referred to as a **batholith** (others may define a batholith as any pluton or association of plutons exceeding 100 km² in surface area). Whereas the geology of the UK contains individual plutons exposed at the surface, some parts of the world have thousands of square kilometres covered almost totally in plutonic rocks. These include the Sierra Nevada mountains of the USA (Box 4.1), and long tracts of Alaska, Peru, Chile and Southeastern Australia. The geological maps of these places are dominated by red – the conventional colour used to depict felsic intrusive rocks. In detail, individual plutons, distinguished on the basis of different mineralogy and with intrusive contacts that cut across structures in older plutons, have built the upper crust in these regions. They record pulses of magmatism on timescales that allowed one plutonic body to have solidified before the next pulse of intruding magma arrived.

Box 4.1   The Sierra Nevada batholith

An extensive part of eastern California is underlain by granite – almost 40 000 square kilometres of it (Figure 4.17a). Its outcrop underpins much of the Sierra Nevada mountain range and the highest peak in the United States outside Alaska, that of Mount Whitney (4420 m above sea level), which is named after the first Director of the California Geological Survey, Josiah Whitney. Much of the area is rugged wilderness with excellent exposures of granite, granodiorite, tonalite and other plutonic rock types making spectacular scenery in the Yosemite, Kings Canyon and Sequoia National Parks (Figure 4.17b). The granite outcrops of the Alabama Hills, outside the town of Lone Pine, have provided backdrops

for generations of classic movies from early Lone Ranger westerns to Star Trek.

The igneous rocks were emplaced through Jurassic and Cretaceous times, with successive intrusions intruding the cores of earlier intrusions. The youngest (late Cretaceous) intrusions form large (1000 km^2) multiple intrusions, exemplified by the Tuolumne Suite whose rocks are exposed throughout Yosemite National Park. They show a roughly concentric arrangement, from tonalite or granodiorite margins to granite centres exposed in dramatic three-dimensional relief thanks to the forces of glacial erosion that have carved valleys more than 1000 m deep (Figure 4.17b).

(a)

key — rest of batholith

— intrusive suite

(b)

Figure 4.17   (a) Map showing the Sierra Nevada batholith and its location within California (inset). The Sonora, Tuolumne, Mono Pass and Mount Whitney intrusive suites are roughly of the same age (late Cretaceous) and were among the last intrusions in the batholith. (b) Tenaya Canyon, Yosemite National Park, exposes the Tuolumne intrusive suite.

The floors of plutons and batholiths occur deep within the crust, and tend not to be exposed. On some map cross-sections you will see a pluton or batholith displayed as a steep-sided body whose base is conveniently assigned to lie below the greatest depth shown on the cross-section. This reflects the fact that, in most cases, what happens at the base is genuinely not known.

When studying igneous intrusions, and indeed almost any aspect of geology, the rocks exposed depend on the depth to which erosion has penetrated into the crust. For instance, removing a few hundred metres from a volcanic system may expose an array of minor intrusions, whereas erosion of several kilometres may expose a pluton which, with further erosion, might reveal a batholith. When studying plutonic rocks, a geologist is seeing the culmination of long, complex and possibly multiple processes of melting, intrusion and crystallisation. It is also likely that many shallow plutons will have fed volcanic eruptions.

Activity 4.2   Plutons on geological maps

In this activity, you will use geological maps to deduce how and when different plutons were emplaced in Britain's geological past.

4.4   Summary of Chapter 4

1   Dykes fill extensional fractures, whereas the emplacement of sills occurs when magma pressure can lift the overlying rocks and allow the magma to spread along a bedding plane or other plane of weakness.

2   Plutons are intrusive igneous bodies a kilometre or more in size. They tend to be coarse grained, whereas minor intrusions are fine or medium grained and have chilled margins.

3   Plutons that were intruded at great depth in the crust tend to be concordant with the structure of their country rock, whereas shallower plutons tend to be discordant and produce metamorphic aureoles.

4   Diapiric rise of plutons is not usually effective at <10 km depth. Plutons emplaced at shallower levels than this are probably fed either from below by dykes or along fault planes, in which case fault movements create the space occupied by the pluton, with some assistance from updoming and stoping.

5   Plutonic rocks can be classified according to the proportions of the felsic minerals quartz, alkali feldspar and plagioclase feldspar.

4.5   Objectives for Chapter 4

Now you have completed this chapter, you should be able to:

4.1 Recognise minor intrusions, plutons and batholiths on geological maps and cross-sections, giving reasons for your diagnosis and suggesting, if appropriate, how they might be related.

4.2 Suggest reasoned explanations for the relationships between an intrusion and its country rock.

4.3 Describe probable emplacement mechanisms for different kinds of intrusion.

4.4 Classify igneous rocks on the basis of their felsic mineral content using the QAP diagram.

Now try the following questions to test your understanding of Chapter 4.

Question 4.5

On the Bedrock UK Maps, find the following igneous intrusions and identify each one as (i) either a pluton or a minor intrusion, (ii) either concordant or discordant, and (iii) specify the age of each as fully as possible on the evidence of the map: (a) Unit *G* between NS (26) 8718 and NY (35) 3388, (b) Unit *CP* centred near SX (20) 6578.

Question 4.6

While visiting a city centre shopping district, you become distracted by an attractive rock embellishing the facade of a shop. The rock's polished surface clearly reveals the coarse-grained interlocking crystalline texture of a plutonic igneous rock. You estimate that the rock has 10% quartz, 60% feldspar and 30% ferromagnesian minerals. There are equal amounts of alkali feldspar and plagioclase feldspar. What rock name should you give it, according to Figure 4.14?

Question 4.7

Complete Table 4.1, which is an aid to distinguishing sills from lava flows, by inserting ticks to indicate the features you might see in cross-section in the field when studying an ancient volcanic area. Insert crosses to indicate features you would definitely not expect to find.

Table 4.1 For use with Question 4.7.

	Sill	**Lava flow**
chilled margin at top		
chilled margin at bottom		
columnar joints		
concordant top with local discordance		
concordant base with local discordance		
rubbly top		
rubbly bottom		
baking of overlying rock immediately above the contact		
baking of underlying rock immediately below the contact		

Chapter 5   The origins of magmas

Volcanic and intrusive rocks are formed when magma solidifies, but why does the molten magma exist in the first place?

5.1   What causes rocks to melt?

You might think that the answer to this question is obvious, and that magma must be a result of heating. In fact, although sufficient heating of a rock will yield magma, heating is not the only, or even the main, cause of melting in the crust and upper mantle.

Let's think about what happens during melting by applying what is known about the reverse process – crystallisation of liquid magma. When a magma solidifies as a result of cooling, different minerals begin to crystallise at different temperatures. In melting, the converse is the case (irrespective of whether melting is caused by heating or some other factor). In any rock consisting of more than one mineral, melting starts by some of the minerals reacting chemically with each other to produce a liquid. As melting proceeds, with for example increasing temperature, each mineral contributes to the melt at a different rate until they are entirely used up. Minerals disappear sequentially in the reverse of the crystallisation sequence. When conditions are such that complete melting is not attained, there are crystals of at least one mineral still surviving. This behaviour is termed **partial melting**, and is typical of the way magmas are produced in the Earth.

For a rock of a particular composition, there is a temperature at which melting will begin and a higher temperature at which melting will be complete. These two temperatures vary according to pressure, and it is conventional to show this relationship on a graph of pressure (P) against temperature (T), with a line joining all points where melting begins (called the **solidus**) and another line joining all the points where the last crystals dissolve and melting is complete (called the **liquidus**). Such a P–T diagram is plotted for mafic rock in Figure 5.1. This is a phase diagram, with the solidus and the liquidus bounding areas in which the material is in different states. It is based on the results of many laboratory experiments in which powdered samples of the rock are held at different high pressures and temperatures, and then quickly returned to room temperature and pressure and examined to see whether they had melted or still contained crystals.

Figure 5.1 is plotted for anhydrous conditions (i.e. no water present), and this is a reasonable approximation to natural conditions because many basaltic magmas have very low water contents, which is why they erupt in effusive or only mildly explosive styles. Phase diagrams such as this are helpful for interpreting the processes that rocks undergo at high pressure and temperature within the Earth. Question 5.1 tests whether you can read and understand the information given by this diagram.

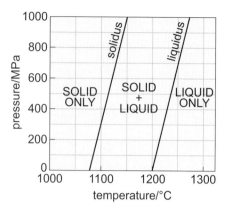

Figure 5.1   Phase diagram showing the solidus and liquidus for a typical anhydrous mafic rock. Pressure is shown in megapascals (1 MPa = 10^6 Pa). (Elsewhere you may find kilobars (kbar) used to denote pressure: 1 kbar (a thousand times atmospheric pressure at sea level) is equal to 100 MPa.)

Question 5.1

(a) Using Figure 5.1, decide whether material of mafic composition consists of solid, liquid, or both, under the following *P–T* conditions: (i) atmospheric pressure (0.1 MPa), 1100 °C; (ii) 1000 MPa, 1100 °C; (iii) 400 MPa, 1150 °C; (iv) 200 MPa, 1250 °C.

(b) If an anhydrous piece of gabbro is heated up keeping the pressure constant at 0.1 MPa, at what temperature does it (i) begin to melt, and (ii) become completely molten?

As the temperature increased from 1080 °C to 1200 °C (in the answer to Question 5.1b), the sample would consist of a decreasing proportion of solid and an increasing proportion of liquid. Just to the right of the solidus it would be mostly solid rock containing a tiny amount of melt, and just before it reached the liquidus it would be mostly melt with a few crystals dispersed in it.

Pressure also has an influence on partial melting.

Question 5.2

With reference to Figure 5.1, describe what would happen if we took an anhydrous piece of basalt at 1000 MPa and 1100 °C and gradually decreased the pressure to 0.1 MPa without changing the temperature.

The phase diagram of basalt shows that partial melting can be brought about by decreasing the pressure without changing the temperature. This process is called **decompression melting** and illustrates that a source of heat is not always necessary to cause melting.

It is also important to realise that when a sample is partially molten, the composition of the liquid is different from the average composition of the remaining solid (because some minerals have contributed disproportionately to the melt), and that both are different to the composition of the starting material. However, the average composition of the melt plus remaining solid must always be identical to the composition of the starting material.

■   Mafic rock consists of olivine, pyroxene and plagioclase feldspar. Of these minerals, olivine has the highest melting temperature (at a given pressure). This means that during partial melting of gabbro, some of the olivine would survive as crystals after everything else had been consumed by melting. If the melt were to be squeezed out, leaving these crystals behind, what name would you give to the rock formed by these crystals? (See Figure 4.13.)

☐   A rock consisting of olivine and little else would be ultramafic in composition and would be described as a peridotite.

Partial melting of mafic material could therefore leave a solid residue of ultramafic composition. On the other hand, the very first melt to form would be intermediate in composition because most of its constituents would be from pyroxene and the least calcium-rich plagioclase. Whether or not all the plagioclase and pyroxene had melted would depend both on their compositions and on the final *P–T* conditions. Broadly speaking, the closer the final conditions are to the solidus, the richer the melt is in silica, but the smaller the compositional change in the solid residue. Conversely, if the final P–T conditions lay only just to the left of the liquidus, the amount of solid residue would be smaller and it would have a more ultramafic composition, whereas the melt would be only slightly richer in silica than the starting material.

This leads us to an important generalisation, as follows:

> When conditions are such as to end up with partial rather than total melting, the melt is richer in silica than the starting material. It is common for partial melting of ultramafic starting material to yield a mafic melt, for partial melting of mafic starting material to yield an intermediate melt, and for partial melting of intermediate starting material to yield a felsic melt.

In addition to pressure and temperature, there is a third factor controlling melting to consider. Figure 5.1 is plotted for anhydrous conditions. This means that there is no water dissolved in the magma, the solid contains no hydrous minerals, and water does not permeate the rock. As soon as water is added to the system, the whole situation changes. Water has the effect of lowering the melting temperatures, so that both the solidus and liquidus are displaced to the left on a P–T phase diagram. Figure 5.2 shows a solidus and liquidus plotted for water-saturated conditions, which are defined as those in which the liquid phase contains as much dissolved water as it can hold. This depends on the pressure, with high pressures allowing more water to be dissolved in the liquid than at low pressures. The amount of water that can be dissolved varies from about 0.1 per cent (by mass) at atmospheric pressure, to many per cent at several hundred MPa. If there is more water present than the magma can hold at a given pressure, the excess water will form a separate phase (usually in the form of vapour bubbles).

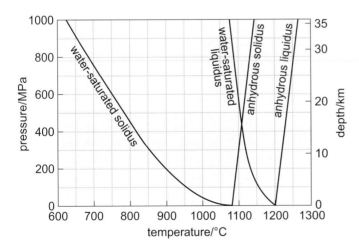

Figure 5.2   Phase diagram for mafic rock showing the solidus and liquidus for water-saturated conditions, in addition to the anhydrous lines shown in Figure 5.1. In hydrous conditions but with insufficient water for saturation, the phase boundaries would lie between the anhydrous and water-saturated extremes. The vertical scale on the right indicates the approximate depth in continental crust corresponding to the pressures on the opposite scale.

On Figure 5.2, the water-saturated phase boundaries occur at lower temperatures than the anhydrous phase boundaries, except at vanishingly small pressure. This is because if there is no confining pressure on the magma, hardly any water can dissolve in the melt, so it is virtually anhydrous.

To explore an important consequence of hydration, you should now try Question 5.3.

Question 5.3

According to Figure 5.2, what would happen if you took anhydrous mafic rock at 1000 MPa and 1100 °C and (without changing P or T) added sufficient water to make the conditions water-saturated?

Thus, melting can occur simply as a result of the addition of water (or indeed any other volatile). Water saturation has an even more extreme effect on the melting behaviour of felsic rock, as shown in Figure 5.3. This draws attention to the fact that the chemical composition of the material influences the melting temperature and pressure: the greater the silica content of a magma, the lower the melting temperature.

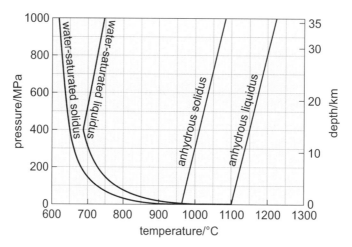

Figure 5.3   Anhydrous and water-saturated phase boundaries for felsic rock. Compare with Figure 5.2. Note that in both cases the water-saturated phase boundaries meet the equivalent anhydrous phase boundaries at zero pressure.

5.2   Crystallisation of magma

Crystallisation can be thought of as the reverse of melting, so the phase diagrams in Figures 5.1 to 5.3 can be used to show the conditions that cause magmas to crystallise.

■ The processes of heating, decompression and addition of water can cause rocks to undergo partial melting, so how likely is it that (i) cooling, (ii) increase in pressure, or (iii) removal of water will cause a magma to crystallise?

□ The phase diagrams indicate that all three of these processes can shift a magmatic liquid from a position on (or even above) its liquidus to a position below the liquidus, and therefore into a stability field where some crystals are present.

You have already seen in Chapter 4 that cooling at the edge of an igneous intrusion causes rapid crystallisation and the formation of small crystals, whereas farther from the contact cooling is slower so fewer but larger crystals grow producing a coarser-grained rock. Although both the fine-grained and coarse-grained rocks will have started at the same temperature and ended at the same temperature, the rate at which the temperature changed will have been different. A fast cooling rate yields finer-grained igneous rock than slower cooling. The reasons for this are best understood after remembering that two separate processes occur. Starting with a liquid, the first step in producing crystals involves the gathering together of a sufficient number of the right atoms to make a tiny speck of a crystal; this is the process of nucleation and the initial speck is called a crystal nucleus (plural nuclei). Crystal nuclei are only a few nanometres in size. The second process is growth of the crystals by the addition of layers of atoms to the nucleus.

For crystals to nucleate or grow, the liquid must be at some temperature below the liquidus temperature. There must also be sufficient time spent at these lower temperatures for nuclei and crystals to achieve a reasonable size, otherwise the liquid will cool so quickly that it quenches to a glass without crystallising. In this case, simply following the temperature path across the stability fields on a phase diagram does not predict the correct outcome. This is because the phase diagram is drawn from the results of experiments that have been carried out over long enough times for the system to reach equilibrium. Rapid cooling can cause disequilibrium, with different outcomes.

With this background, we can consider the texture of an igneous rock in terms of the time spent below the liquidus and the amount of cooling beneath the liquidus. The difference between the liquidus temperature and the actual temperature is called the **undercooling**, often denoted by the symbol ΔT (pronounced 'delta T'). The rates at which nuclei form (measured in terms of the number of nuclei forming in a cubic millimetre per second) and at which crystals grow (lengthen) depend on ΔT. You might expect that these rates increase with increasing ΔT, but this is only partly true. At large undercoolings, the atoms in the liquid move very slowly within the melt structure, for similar reasons as to why viscosity increases with falling temperature. Experiments have shown that at large undercoolings, nucleation and growth rates are both small. Graphs in which the nucleation rate and growth rate are plotted against ΔT illustrate these features (Figure 5.4). Particularly important features are that the peak growth rate occurs at lower undercooling than the peak nucleation rate, and that at large undercoolings, both the nucleation and growth rates are very small.

Based on Figure 5.4, it is apparent that magmas may crystallise under three regimes:

- very low undercooling, where nucleation rate is low but growth rate is high
- moderate undercooling, where nucleation rate is high but growth rate is low
- high undercooling, where both nucleation and growth rates are very low.

These three regimes lead to different textures.

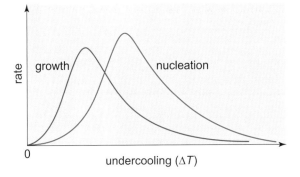

Figure 5.4   Schematic graphs of nucleation rate and crystal growth rate as functions of undercooling. There are no scales on the axes, partly because crystal nucleation and growth rates are measured in different units so the relative heights of the two curves are not significant, but note that the peak growth rate occurs at lower ΔT than the peak nucleation rate.

Question 5.4

Look at the three sketches of igneous rock textures in Figure 5.5 and decide which one was produced under each of the nucleation and growth rate conditions listed in Table 5.1.

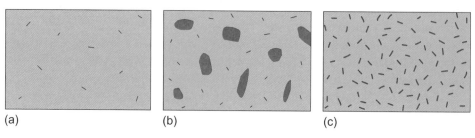

(a)   (b)   (c)

Figure 5.5   Sketches of three examples of igneous rock textures showing crystals (grey) embedded in glass (pink). For use with Question 5.4.

Table 5.1   For use with Question 5.4.

Nucleation rate and growth rate	Undercooling	Example in Figure 5.5
nucleation rate low, growth rate high	low	
nucleation rate high, growth rate low	moderate	
nucleation rate very low, growth rate very low	high	

We can now consider the circumstances that determine the size of undercooling a magma experiences at the time of crystallisation, and the cause of the undercooling. The simplest case is, of course, where hot magma is intruded or erupted into a colder environment, and cooling proceeds from the outside of the magma inwards. This is the situation that applies to an intruded sill (Figure 4.6) or a pillow lava. The greatest undercooling is imposed at the contact so here the finest-grained rock is formed. At the edge of a pillow lava, where basalt at a typical temperature of 1150 °C is brought into contact with cold water, the undercooling can be so large that no nuclei form and the liquid is quenched to a glass in the outermost 1 or 2 cm of the pillow. In the interior, the temperature changes slowly and the magma spends a longer time at temperatures not far below the liquidus (low undercooling), allowing phenocrysts to grow.

But cooling is not the only way to achieve sub-liquidus conditions and cause crystallisation within a magma. An alternative process to cooling is to remove water from a magma by decompressing it.

■   Why can decompression of a hydrous magma lead to undercooling (assuming that the temperature does not change during this process)?

☐   Once the pressure falls to a small enough value, the magma becomes saturated in water and bubbles will form with further decompression, removing water from the liquid. At a given pressure, the liquidus temperature of water-poor magma is higher than that of water-rich magma (Figures 5.2 and 5.3), so the magma will become undercooled once it starts to exsolve water.

The eruption of magma has the potential to induce crystallisation by virtue of it losing water, but how effective this will be depends on the amount of time available. A rapidly rising (rapidly decompressing) magma may not have sufficient time for nuclei or crystals to grow, in which case it will erupt as a supercooled liquid. Alternatively, more slowly erupting magma may grow large amounts of tiny crystals. Interestingly, the growth of many crystals causes the magma to increase its viscosity. This in turn causes the magma to flow more slowly driving yet more crystallisation, which is an example of a feedback process that might result in a change in the character of the eruption from explosive to effusive, or even cause it to stop if the magma became so crystal-rich as to become immobile.

5.3   Chemical composition of magma

Igneous rocks and magmas come in a wide range of compositions, from silica-poor ultramafic rocks to silica-rich felsic rocks. How is this range explained? Partial melting yields liquids which are different in chemical composition from the rock that was melted. This means that the composition of the starting material and the amount of partial melting will both determine the composition of the resulting liquid magma, known as a **primary magma**. Once a primary magma has been formed by partial melting it will normally migrate upwards, intruding into colder rocks and start to cool down, which has the effect of inducing crystallisation. Crystallisation also has the effect of changing the chemical composition of the liquid part of a magma because the early-formed crystals have a different chemical composition from the primary magma. In the case of a magma that is slowly cooling in a subterranean magma chamber, phenocrysts will grow in the magma. If these phenocrysts do not move while further cooling and crystallisation of the magma occurs, then by the time the magma has completely solidified it will still have the same composition as the original magma.

■ How will the composition be affected if the first-formed crystals become separated from the melt?

☐ The remaining melt will be more felsic than the original melt, whereas the separated crystals will form a rock less felsic (more mafic) than the original melt.

Such physical separation of the crystal fraction from the melt fraction in a partly crystallised magma is a very important way in which the composition of a magma can change. The process is termed **fractional crystallisation**. It can occur by the residual liquid being squeezed out of the space between the crystals, or by crystal settling if magma is stored for a long period in a magma chamber. The result of fractional crystallisation is a liquid whose chemical composition is more felsic than the original melt, and a body of separated crystals called a **cumulate**. So some igneous rocks are cumulates, and include certain types of layers in layered intrusions (e.g. Figure 4.16b).

■ What types of minerals would you expect to form cumulate rocks?

☐ Cumulates form from early crystallised minerals, which are those that crystallise at high temperature. These are mafic ferromagnesian minerals, such as olivine and pyroxene, and calcium-rich plagioclase feldspar.

Another way in which magma composition can change is if the magma has enough heat to melt some of the rock it passes through. If this new melt is able to mix into the main magma body, then this will develop a new composition that is a weighted average of its initial composition and that of the new material that has been mixed in. This process is described as **assimilation**.

■   Do you think it likely that a rising body of felsic magma would assimilate mafic rocks through which it was rising? (Refer to Figures 5.2 and 5.3.)

☐   Mafic rocks begin to melt at a much higher temperature than felsic rocks, so a felsic magma would not normally be hot enough to melt a mafic rock.

The converse is more common, in that a rising body of mafic magma can assimilate rocks of intermediate and felsic composition, and a rising body of intermediate magma can assimilate rocks of felsic composition. In the process, the magma becomes more felsic than its initial composition. Both fractional crystallisation and assimilation therefore cause the silica content of a magma to increase, and the magma is said to evolve in composition. In relative terms, felsic magmas are said to be more evolved than intermediate magmas, which are more evolved than mafic magmas.

5.4   Magmatism and plate tectonics

We conclude our discussion of igneous processes with a look at the relationships between the origin of magmas, magma evolution and the broader workings of the Earth's interior, particularly in relation to plate tectonics.

5.4.1   Magmatism at divergent plate boundaries

Mafic magma, which abounds at divergent (constructive) plate boundaries, is what you would expect to find as a consequence of partial melting of the ultramafic mantle, and the mantle is the only potential melt source in such a setting. It is also natural that magma should escape upwards, because mafic magma is less dense than ultramafic rock. However, why the mantle should partially melt is not quite so obvious.

At mid-ocean ridges, two oceanic plates diverge and asthenospheric mantle is drawn upwards from beneath the plate boundary at a rate sufficient to plug the gap (Figure 2.8).

■   What will happen to the pressure experienced by this asthenospheric mantle as it is drawn upwards beneath the divergent plate boundary?

☐   The pressure must drop as the depth decreases.

To see the implications of this, look at the P–T phase diagram for material of ultramafic composition (Figure 5.6). The mantle beneath mid-ocean ridges is typical of normal upper mantle and has a very low water content, so it is the anhydrous solidus that is relevant in controlling partial melting in this setting. The diagram shows a likely path followed as asthenosphere is drawn upwards beneath a divergent plate boundary.

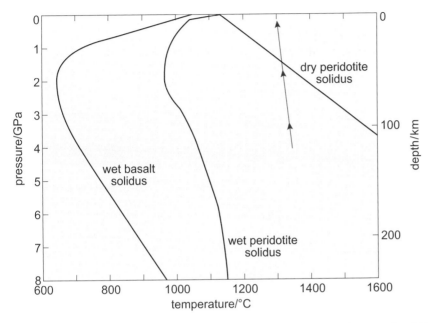

Figure 5.6 The path of changing pressure and temperature (arrowed red line) followed by mantle upwelling beneath a divergent plate boundary, and the locations of the solidus of dry (anhydrous) peridotite, wet peridotite and wet basalt. Note that this diagram has been drawn with pressure increasing down the vertical axis, and with pressure shown in gigapascals (1 GPa = 1000 MPa).

■ According to Figure 5.6, what will happen to upwelling anhydrous mantle when it reaches a depth of about 45 km?

☐ This is the depth at which it crosses the anhydrous (dry) solidus, so it will begin to melt.

The exact depth at which partial melting begins below a particular divergent plate boundary depends on factors such as the starting temperature, and the speed and cooling rate of the upwelling asthenosphere. However, the important point is that although upwelling mantle cools slightly as it rises, the drop in pressure is such that it can rarely avoid crossing the solidus into the partially molten field. Mafic magma at divergent plate boundaries is therefore generated by partial melting of the ultramafic mantle entirely as a result of decompression melting, a process you met in Question 5.2.

Sea-floor spreading induces partial melting of the underlying mantle, releasing basaltic magma that rises to the surface. 'Fieldwork' at mid-ocean ridges involves submersible dives and has found basaltic lava flows that are too young to have become covered in the slowly accumulating layer of deep-sea sediment that blankets the ocean floor at greater distances from the ridge. Pillow lavas are particularly common (Figure 5.7). To find out about deeper portions of the oceanic crust geologists must rely on seismic studies, a limited amount of drilling, and sampling (by submersibles and dredging) at places where deeper levels are exposed (e.g. at transform faults and fracture zones).

Figure 5.7   Pillow lavas on the Mid-Atlantic Ridge, 3150 m below sea level. Each large 'pillow' is about 0.5 to 1 m in diameter.

The picture that emerges from combining all these lines of study is that lava flows make up roughly the top half kilometre of the oceanic crust, which is itself about 7 km thick on average. The lowest part of the oceanic crust is a coarse-grained igneous rock containing predominantly pyroxene and plagioclase feldspar.

■ What rock name would you give to this rock type?

☐ According to Figure 4.13 this is gabbro, the coarse-grained (i.e. plutonic) equivalent of basalt.

The gabbro layer is believed to form by slow solidification of intrusions of basalt magma beneath a mid-ocean ridge (hence its large grain size). Fractional crystallisation of olivine, plagioclase and pyroxene during slow cooling of the gabbro magma chamber can produce relatively evolved basalt magma that may spill out onto the ocean floor and erupt as pillow lava. Lying above the gabbro and below the lavas is a layer about 1 km thick consisting of nothing but parallel dykes of medium-grained dolerite. Each dyke has been intruded up the middle of a previous dyke or up the contact between two previous dykes. This layer is referred to as a **sheeted dyke complex**. The three-layer structure of oceanic crust, summarised in Figure 5.8, can be well displayed where segments of old oceanic

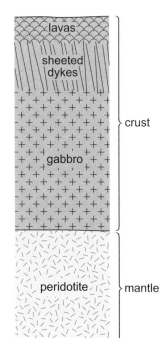

Figure 5.8   The three-fold division of igneous oceanic crust into basalt lavas, dolerite dykes and gabbro.

crust and upper mantle tens or even hundreds of kilometres across have been thrust up onto continental regions, as found for example in Oman. Such slabs of ocean floor are known as an **ophiolite** (pronounced 'oaf-ee-oh-lite') or an ophiolite complex (Box 5.1).

Igneous processes at divergent plate boundaries drive other processes where seawater is drawn down into the hot young crust, becoming heated and then rising back towards the sea floor in a hydrothermal circulation system. The hot water reacts with some of the minerals in the crust and changes their chemistry (in particular, olivine becomes altered to serpentine and much of the pyroxene is replaced by the hydrous mineral amphibole) and veins of metal-rich minerals may be deposited.

Box 5.1 The Troodos Mountains, Cyprus

The geology of Cyprus, in the eastern Mediterranean Sea, is largely a combination of late Mesozoic and Cenozoic sedimentary rocks and slightly older igneous rocks. It is the igneous rocks that form the main mountain range – the Troodos Mountains – and they have been dated at 90–92 Ma. The igneous rock types are basaltic pillow lavas, parallel dolerite dykes packed so tightly that entire tracts of land are composed of metre-scale dykes, gabbro, and coarse-grained olivine–pyroxene rocks. The mineral content of the latter rock type defines them as varieties of the ultramafic rock type peridotite. Some of the peridotites have been heavily altered by reaction with hot water, turning olivine into serpentine; such rocks are known as serpentinite – an allusion to the impression that the rocks look like snake skin. The simplified geological map in Figure 5.9a shows that these are arranged in an almost concentric fashion over an area about 90 km long and 40 km wide. Pillow lavas are exposed almost all the way around the fringe of the outcrop. From the geological map, it is apparent that the pillow lavas are overlain by younger sediments. Copper sulfide mineralisation is widespread within the pillow lavas and has been exploited for more than 4000 years; it is no surprise that Cyprus derives from the Greek word for copper (*cupros*). Beneath the pillow lavas lie the dyke rocks (forming a sheeted dyke complex because this rock unit is composed of sheet-like intrusions; Figure 5.9b) and these pass into gabbro, which is in turn in contact with the peridotites and serpentinites.

The origin of the igneous rocks is evident from their geology: sea-floor eruption of pillow lava, shallow magma transport in dykes, plutonic crystallisation of gabbro, and peridotite typical of upper mantle rocks. Viewed as a whole they represent a cross-section through mafic igneous crust, from the mantle roots up through to the intrusive magma storage regions and transport routes that once generated new oceanic crust at a mid-ocean ridge.

The Troodos ophiolite was the first such example to be interpreted in terms of plate tectonic processes once a major mapping project carried out by Greek and British geologists during the 1960s led to the realisation that the Troodos rocks matched the rock types of the oceanic crust, despite now being well and truly part of continental lithosphere. A key scientist in these discoveries was Professor Ian Gass, the first professor of Earth Sciences at The Open University.

The outcrop pattern on Figure 5.9a and the inclined dip of the sheeted dykes (Figure 5.9b) show that the rocks are no longer in their original horizontal orientation, and this is because of deformation that occurred while the peridotites were being altered to serpentinite. This chemical reaction introduces water into the mineral structure, causing a large expansion of volume (and decrease in density) which causes the serpentinite to swell upwards, shouldering aside overlying rock and intruding into it. The Troodos ophiolite is an example of oceanic crust that has become sufficiently buoyant to avoid subduction.

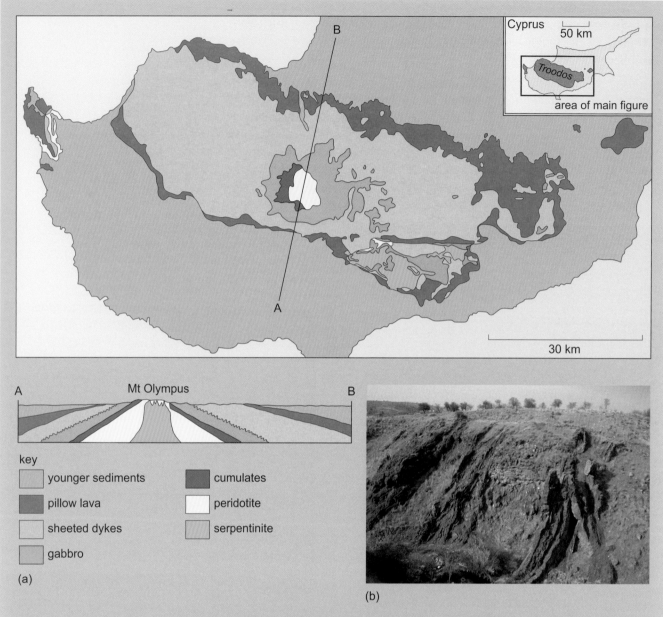

key

- younger sediments
- pillow lava
- sheeted dykes
- gabbro
- cumulates
- peridotite
- serpentinite

(a)

(b)

Figure 5.9 (a) Simplified geological map of the Troodos Mountains, with insets showing their location on Cyprus and a north–south cross-section. (b) Sheeted dykes intruding pillow lavas, Klirou, Cyprus. Dykes are 1–2 m wide.

5.4.2 Magmatism at hot spots

Basaltic volcanism abounds on the Big Island of Hawaii, in the middle of the Pacific Ocean, thousands of kilometres from any plate boundary. Indeed, the whole island is composed of basalt. This is a **hot spot**, where large volumes of basalt erupt in isolation from plate boundaries. Volcanism here, and at other **intraplate volcanoes** (or **within-plate volcanoes**) is best explained by decompression melting of plumes of unusually hot mantle rising buoyantly from great depth, possibly from as deep as the core–mantle boundary. On approaching the lithosphere, the material in the plume is at a sufficiently low pressure (and high temperature) to partially melt, releasing basaltic magma that erupts through the overlying lithosphere. In most cases the lithosphere is itself moving, but in a horizontal direction as part of plate tectonics. As the lithosphere moves over the hot spot, a chain of extinct volcanoes is left behind on the surface. Active volcanoes lie above the hot spot but progressively older extinct volcanoes lie at increasing distances away, as classically illustrated in the 6000 km-long Hawaii–Emperor chain of islands and seamounts (sub-sea volcanoes) (Figure 5.10).

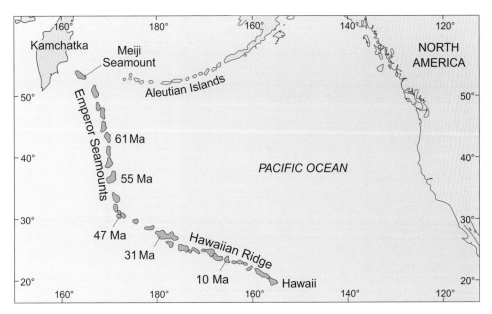

Figure 5.10   The islands and seamounts of the Hawaii–Emperor chain (brown), showing an age progression (in millions of years) away from the active hot spot beneath Hawaii.

If mantle plumes are initiated deep in the mantle, then they can potentially appear at the surface under any region of the Earth, such as in the middle of an oceanic plate in the case of Hawaii. Unusually intense volcanism in Iceland has built this part of the Mid-Atlantic Ridge up above sea level and requires unusually productive partial melting, which is attributed to a hot spot coinciding with a divergent plate boundary. Other examples of hot spots occur within continental lithosphere, such as in the southern Sahara, where about two dozen volcanic centres, some with summit calderas, make up the Tibesti Mountains, rising to 3400 m. There is no obvious linear chain of extinct volcanoes here because the African Plate is hardly moving relative to the underlying mantle plume. Similarly, the dominantly basaltic Cape Verde and Canary Islands (Box 5.2) overlie hot spots on the oceanic portion of the African Plate. A continental hot spot track is, however, found on the faster-moving North American Plate (Figure 5.11), where a series of large rhyolitic ignimbrite calderas are interpreted as resulting from assimilation and melting of continental crust as it passes over a hot spot. This hot spot currently underlies the Yellowstone caldera. The older calderas lie to the southwest and become younger towards Yellowstone, indicating that the North American Plate is moving southwest relative to the hot spot.

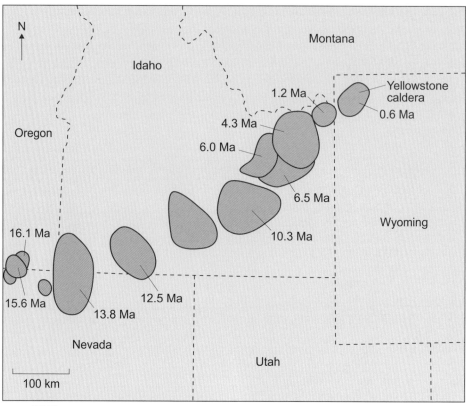

Figure 5.11   Map of the northwestern USA, showing the locations and ages of rhyolite calderas that formed as the North American Plate moved over the Yellowstone hot spot.

Box 5.2 Volcanoes of Tenerife, Canary Islands

Tenerife, the largest of the Canary Islands, is the third largest volcanic ocean island on Earth by volume (Figure 5.12a). It is dominated by El Teide (or Mount Teide), a World Heritage Site, which forms part of the Parque Naçional del Teide. At 3718 m above sea level, and approximately 7500 m above the floor of the Atlantic Ocean, Teide is the highest mountain in Spain, and the highest point in the Atlantic Ocean. The active, but dormant stratovolcanoes of Teide and its sister peak Pico Viejo ('Old Peak', even though it's actually younger!) form the most recent central volcanic complex on Tenerife.

Teide and Pico Viejo are located in a 16 km × 19 km depression known as the Las Cañadas caldera that was formed about 200 000 years ago, although the mechanism (catastrophic eruption or collapse) is still a matter of contention among researchers. The southern part of the caldera wall provides a cross-section through the older volcano (Figure 5.12b). Much younger eruptions of felsic magmas produced pyroclastic flows and fall deposits, including those known as Bandas del Sur, that are well exposed along the south coast of the island.

Eruptions from the northwest and northeast rift zones are thought to be fed from dykes extending laterally from beneath the summit. Eruptions on Tenerife in historical times include lava flows from near the summit of Pico Viejo in 1798 and the 1909 eruption of a small strombolian mafic cone with associated lava flows and black scoria fall from vents on the northwest rift zone (Montaña Chinyero). Teide currently has active fumaroles at the summit, which release hydrogen sulfide and other gases.

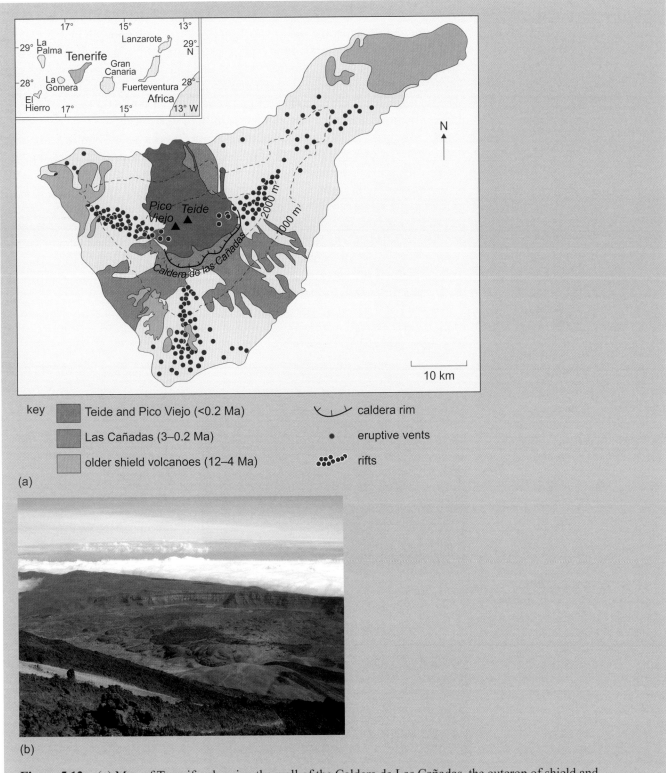

key

■ Teide and Pico Viejo (<0.2 Ma)

■ Las Cañadas (3–0.2 Ma)

□ older shield volcanoes (12–4 Ma)

⌣⌣ caldera rim

● eruptive vents

⦂⦂⦂ rifts

(a)

(b)

Figure 5.12   (a) Map of Tenerife, showing the wall of the Caldera de Las Cañadas, the outcrop of shield and stratovolcanoes, and the distribution of vents along the northeastern, northwestern and southern rift zones and their associated lavas (green). (b) Photograph from the flanks of Teide, showing a'a lava flows in the foreground and the floor and southeastern part of the Caldera de Las Cañadas. Note the layers of lava exposed within the caldera wall. Lava and pyroclastic deposits from the 2000-year-old Montaña Blanca eruption occur on the caldera floor.

5.4.3   Flood basalts

At around the time that the oldest rhyolitic caldera associated with the Yellowstone hot spot was forming (about 16 million years ago), a huge burst of basaltic volcanism occurred to the north, in the states of Oregon and Washington. The most recent investigations and mapping indicate that some 230 000 km^3 of basaltic lava inundated 200 000 km^2 in a series of enormous eruptions fed from a dyke swarm in the space of about half a million years. It built a volcanic pile about 1 km thick, known as the Columbia River Basalt Group (Figure 5.13a). It is the youngest example on Earth of a **flood basalt province** – characteristically large by the scale of present-day volcanic systems and typically involving about one million cubic kilometres of rather monotonous basalt lavas emplaced in a timespan of about one million years or less (a long-term output rate ten times that of Hawaii). Many flood basalt provinces are to be found on the continents, but they can also be erupted onto the ocean floor. This is how the Ontong–Java Plateau in the western Pacific (about 120 million years ago) was formed. The Columbia River Basalt Group is actually on the 'small' side in comparison with the Deccan Traps basalts that cover a large part of western India (Figure 5.13b). It took only half a million years, about 65 million years ago, for more than one million km^3 of basalt magma to be erupted while India and the Seychelles were starting to split apart.

Figure 5.13   Examples of flood basalts. (a) Basalts of the Columbia River Province exposed in a canyon of the Columbia River at Palouse Falls, Washington, USA. Total thickness of lavas here is 115 m. Note vertical jointing in lavas. (b) Map showing the present-day extent of the Deccan Traps continental flood basalts in India. The average total thickness is about 1 km. They cover an area of 0.5×10^6 km^2 today, but prior to erosion, and taking into account what is below sea level, their original extent may have been 1.5×10^6 km^2. (c) Flat-lying lavas on western Mull, dating from about 60 million years ago, in which individual flow units have eroded to produce a stepped topography. Reproduced with the permission of the British Geological Survey © NERC. All rights Reserved.

(a)

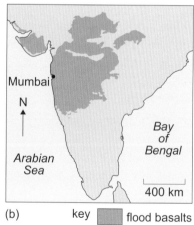

(b)   key   flood basalts

(c)

Continental flood basalts were also erupting during the rifting of the British Isles from Greenland in the early Palaeogene, heralding the opening of the North Atlantic Ocean along the Mid-Atlantic Ridge and vigorous magmatism at the Iceland hot spot. The remains of these flood basalts (the North Atlantic Igneous Province) are to be found in Greenland, the Faeroes, and on the fringes of western Scotland and Northern Ireland (Figure 5.13c). Included within this province are volcanic and plutonic centres on the islands of Mull, Rum and Skye, some of whose shallow and plutonic parts were examined in Chapter 4 (e.g. Figure 4.16).

Flood basalts seem to owe their origin to the arrival of anomalously hot mantle near the Earth's surface, sometimes coupled with stretching and splitting of continental lithosphere. Both processes aid the process of decompression melting that is required to yield exceptionally large quantities of basalt. A useful working hypothesis is that the first burst of activity of the plume is the most volcanically productive because the top of the plume carries the hottest material whereas later plume material, whilst still anomalously hot, is somewhat cooler and therefore produces less melt. This explanation accounts for a long-lived period (tens of millions of years) of hot spot volcanism that produces a hot spot trail (such as the Hawaii–Emperor chain, the Yellowstone chain, and the Iceland hot spot) following the rapid emplacement of a flood basalt province. The exact mechanisms of these processes are not yet resolved, and the details may vary between provinces. Some researchers argue that the available geological evidence is inconsistent with anomalously high temperatures in the mantle being the cause of flood basalt provinces and other voluminous centres of basaltic volcanism that lie far from plate boundaries. Instead, they argue that localised compositional variations in the upper mantle are responsible for regions of anomalously high degrees of partial melting.

Whatever their exact cause(s), flood basalts seem to be erupted at random intervals at random places on the globe, and are among the clearest demonstrations that the Earth's interior is not in a monotonously uniform steady state, but undergoes episodes of enhanced activity. The lava erupted in a large flood basalt province would clearly have devastating local effects, but the amount of volcanic gas liberated over a comparatively short period could also have significant effects on global climate and other environmental systems. The environmental disturbances caused by the eruption of at least two flood basalt provinces – the Deccan Traps (65 Ma) and the Siberian Traps (250 Ma) – may have contributed to the mass extinction events at the end of the Cretaceous and Permian Periods, respectively. Fortunately for us and many other species, there are no indications that eruption of another suite of flood basalts is imminent.

5.4.4   Magmatism at subduction zones

Most of the world's on-land volcanoes occur at convergent plate boundaries on the edges of continents and in oceanic island arcs where andesitic stratovolcanoes of the kind described in Section 3.6.2 predominate. In this subsection, we shall examine how subduction leads to volcanism. Once again, phase diagrams hold the key.

Look back at Figure 2.5. This shows oceanic lithosphere being subducted, and volcanoes situated above the subduction zone. Clearly, melt must be being generated somewhere below the volcanoes. Consider what happens to the downgoing plate, usually referred to as the **slab**, during subduction.

◼ What will happen to the pressure and temperature in the slab?

☐ The pressure will increase according to depth. As it gets deeper, the slab comes into contact with progressively deeper and warmer parts of the mantle. Heat will be conducted into the slab, causing it to warm up.

Pressure is transmitted instantaneously, so the pressure in the slab must always be the equilibrium pressure for the depth, but the rate at which heat can be conducted into the slab is limited by its thermal conductivity. Therefore, the slab will warm up slowly, at a rate that will depend on how quickly the slab is subducting and how cold it was when it entered the subduction zone. These parameters certainly vary between subduction zones because not all plates move at the same speed and old oceanic lithosphere is colder than younger oceanic lithosphere.

◼ What combination of relative age and plate speed will favour the slab reaching high temperatures at shallow depth?

☐ Young (warm) lithosphere subducting slowly will have been heated to a higher temperature by the time it reaches a given depth than will old (cold) lithosphere.

Mathematical models of the temperatures within subduction zones suggest that only in the case of the youngest lithosphere will the subducted slab reach its solidus because of heat conducted into it from the hot mantle it encounters. This limitation may explain why lavas with specific intermediate compositions of the sort expected from the partial melting of basalt at high pressure are rare. They are found in the few arcs (such as parts of the Aleutians and southern Chile) where the plate age and speed would be appropriate for the slab to melt. However, the simple geological observation that basalt magmas are erupted from volcanoes in all subduction zones means that partial melting of the basaltic slab cannot be the sole, or even dominant, process for generating magmas in this tectonic setting. To produce basalt, ultramafic peridotite must be partially melted, and this is the rock type that makes up the **mantle wedge** lying above the subducted slab. But the solidus of peridotite is at even higher temperatures than that of basalt, and the wedge is cooled by the cold slab plunging into it. This does not seem to be a situation that could cause the mantle wedge to melt. To understand how basaltic magmas are generated within arcs requires closer consideration of the nature and fate of the slab.

Although oceanic crust is created by partial melting from an anhydrous source (as you saw in Section 5.4.1), by the time it is subducted anhydrous conditions no longer apply.

■   What are the reasons for this?

☐   The most obvious reasons are that a subduction zone begins under water, so the upper part of the slab must be wet. Seawater will permeate cracks and fissures in the oceanic crust, and the veneer of sediments overlying the lavas (which may be subducted with the slab) will also be wet. As you saw in Section 5.4.1, much of the original pyroxene in the crust of the slab will have been hydrothermally replaced by amphibole, which is a hydrous mineral.

The descending slab is made largely of hydrous rocks, as a result of reactions between anhydrous oceanic lithosphere, produced at a divergent plate boundary, and seawater at low pressures. Subduction has the effect of placing these rocks in a high-pressure environment with the result that they experience metamorphic reactions that change a hydrous mineral assemblage to an anhydrous mineral assemblage. These reactions are known as **dehydration reactions** and they release water from the slab. For example amphibole-rich rock (amphibolite) dehydrates to **eclogite**, a rock dominated by pyroxene and garnet. The water leaves the slab and enters the mantle wedge, turning initially dry peridotite into wet peridotite.

The significance of this stems from the fact that water lowers the melting temperature (solidus) of rock (Figures 5.2, 5.3 and 5.6).

Question 5.5

The mantle wedge beneath volcanic arcs lies between the Moho, at around 35 km depth (1000 MPa, or 1 GPa), and the top of the slab at about 100 km depth (3000 MPa, or 3 GPa). If the temperature in this region ranges from 1000 to 1300 °C, what can you deduce from Figure 5.6 about the ability of this region to yield partial melts?

Water not only plays a critical role in the genesis of magmas at subduction zones, but it also becomes an important component of the magma, with arc magmas containing several weight per cent water. This is the driving force behind the explosivity of many subduction zone volcanoes.

Magmatism in arcs (and other tectonic settings) stems from the production of basalt by partial melting of mantle peridotite. The crust of the overriding plate has no direct role in the process. However, magmas of intermediate and felsic composition are also intruded and erupted at subduction zones, granites and felsic ignimbrites being particularly notable at continental arcs. This is because fractional crystallisation and assimilation (Section 5.3) can cause the silica content of magma to increase during ascent. The full interplay of processes is complex, and varies between volcanoes and through the lifetime of a single volcano, but the essential story is summarised in Figure 5.14.

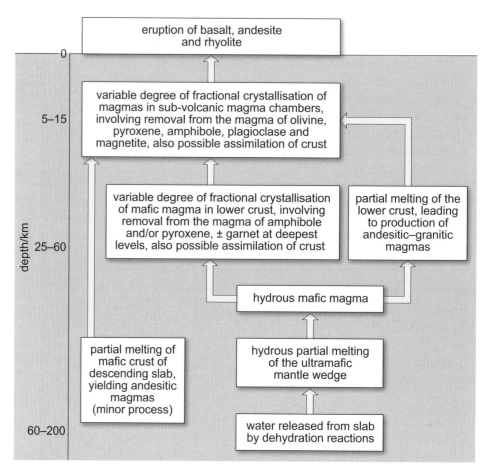

Figure 5.14 Possible processes involved in the generation and evolution of magmas at subduction zones.

Question 5.6

Summarise the difference between magma generation and evolution at subduction zones (Figure 5.14) and at mid-ocean ridges, paying attention to: (a) depth and process of magma generation by partial melting; (b) depth and degree of fractional crystallisation and crustal assimilation; and (c) the composition of the erupted magmas.

5.4.5 Magmatism associated with continental collision

There remains only continental collision to complete our survey of the settings for igneous processes. Figure 2.6b showed that subduction stops after two continents have collided, at which time the magma-generating processes described in Figure 5.14 must also terminate. However, it is found that plutons, especially granites, are emplaced in continental collision zones for several tens of millions of years after a collision. The Lake District batholith is an example of this type, having been intruded in the aftermath of a collision that united England and Scotland in the Silurian.

Geochemical arguments can be used to demonstrate that there is a substantial contribution from the mantle in some post-collision granite magmas, which is not very well understood. Other granite magmas can be explained more simply by considering what collision does to the crust. Compressional deformation of the leading edge of the overriding continental crust combined with partial subduction of the leading edge of the continental crust on the subducted plate means that the crust near the suture zone is much thicker than before collision (Section 2.3.1). For example, the continental crust below the Himalaya (the site of a suture zone between India and the main mass of Asia) is about 70–90 km thick, or about twice the average continental thickness.

■ What are the relative contributions of continental crust and mantle to radiogenic heating within the Earth?

☐ Volume for volume, continental crust generates heat at a rate orders of magnitude greater than the mantle (see Section 2.4).

Deep in the thickened continental crust of a collision zone radiogenic heating will slowly raise the temperature, insulated by a thick layer of crust that is itself producing radiogenic heat by the decay of K, U and Th. Gradually, the temperature may become hot enough for the deep crustal rocks to melt.

■ Which compositional attributes will lead to a rock having a low solidus temperature at a given pressure?

☐ Rocks that are rich in felsic components, that are made from several minerals, and which contain hydrous minerals such as amphibole and mica, have lower solidus temperatures than rocks without these attributes.

The crustal rock types with the lowest solidus temperatures are mudstones and their metamorphic equivalents. Reaction of muscovite and quartz in these rocks will yield a liquid with the composition of granite and an unmelted solid residue. Together these form a migmatite and, if the liquid can segregate from its residue, give rise to granite intrusions (Figure 4.9a). Many of the granites in collision zones were formed by the melting of muddy sediments and other less silica-rich crustal rocks (average continental crust is intermediate in composition, see Section 2.2.1) as a result of the high rates of radiogenic heating within the thickened crust.

You will take a closer look at how heat flow is affected by crustal thickening in Chapter 6, but for now perhaps you will be relieved to have at last found a melting mechanism that relies on heat rather than pressure changes! You have also reached the interface between igneous and metamorphic processes.

Activity 5.1   Igneous rocks and igneous processes

In this activity, you will relate igneous rock specimens included in the Digital Kit and Virtual Microscope to the processes that formed them.

5.5 Summary of Chapter 5

1   Any silicate rock consisting of more than one mineral melts over a range of temperatures and pressures. In a *P–T* phase diagram, the solidus and liquidus separate stability fields of solid only, solid (crystals) + liquid, and liquid only.

2   Under hydrous conditions, or in the presence of any other volatile, both the solidus and liquidus plot at lower temperatures for a given pressure (except for zero pressure).

3   Partial melting of silicate rocks yields a melt that is more felsic than the starting material. This can become separated from the remaining crystals, leaving these behind as a residue that is less felsic (more mafic) than the starting material.

4   Magmas tend to rise because they are less dense than solid rock.

5   A magma can evolve to a more felsic composition by fractional crystallisation (if the crystals are removed) as it cools within a magma chamber, or by assimilating crustal rocks of more felsic composition.

6   At divergent plate boundaries and intraplate volcanoes, mafic magma is generated by decompression melting of anhydrous upwelling mantle.

7   At subduction zones, partial melting occurs in hydrous conditions. Water escaping from the slab hydrates the mantle wedge to produce partial melts of mafic (basalt) composition. Melts may become more felsic during ascent, mainly as a result of fractional crystallisation but perhaps also through crustal assimilation.

8   The rate of radiogenic heat production in continental crust that has been thickened as a result of continent–continent collision can cause sufficiently high temperatures for magmas of mainly felsic composition to be produced by partial melting within the crust.

5.6 Objectives for Chapter 5

Now you have completed this chapter, you should be able to:

5.1 Use *P–T* phase diagrams to explain how melting and crystallisation may occur as a result of changes in pressure, temperature or water content.

5.2 Explain how the composition of igneous rocks depends on a combination of melting, crystallisation and assimilation of country rocks.

5.3 Describe the most common types of igneous activity at divergent and convergent plate boundaries and in intraplate settings, and explain how the magma is generated and how its composition may evolve during ascent.

5.4 State the igneous rock types most likely to form in each plate tectonic setting, but be aware of possible exceptions.

Now try the following question to test your understanding of Chapter 5.

Question 5.7

In continental areas, temperature usually rises with depth at a rate of 20–40 °C km^{-1}. Suppose the rate in a particular region is 30 °C km^{-1}.

(a) What would you expect the temperature to be at a depth of 30 km? (Assume that the surface temperature is 10 °C.)

(b) Plot your answer to part (a) on Figure 5.3, using the depth scale on the right of the diagram, and label it point A. If conditions are anhydrous, explain whether the physical state of felsic material will be solid, liquid, or both.

(c) Suppose that sufficient water is introduced to saturate the system. Explain what the physical state of the felsic material will be now.

(d) Suppose the resulting water-saturated magma rises towards the surface, and that it loses heat at an average rate of 5 °C km^{-1}. Plot a line starting from point A representing the changing P and T followed by magma cooling at this rate as it rises, and with reference to this line describe what will happen. Speculate on whether or not any magma will reach the surface.

Chapter 6   Metamorphism

6.1   Introduction

As described in Section 2.4, the Earth's crust is heated both by primordial heat and by radiogenic heat from decay of the heat-producing isotopes of K, U and Th. This is why temperatures increase with depth inside the Earth. Granites in continental collision zones are formed by melting crustal rocks, but heating the Earth's crust does not always result in melting. As rocks become hotter, their constituent minerals recrystallise and form different minerals that are stable at higher temperatures. Deeper in the Earth's crust (or mantle), the mass of the overlying rocks results in higher pressures imposed on the rocks and new minerals form that are stable at these pressures. Moreover, the shapes and alignments of minerals change in response to increasing pressures as rocks reorganise themselves to take up smaller volumes or different shapes. The changes in the mineralogy and in the orientation of minerals that make up a rock when subjected to changes in temperature and/or pressure are known as metamorphism.

Broadly, the study of metamorphic rocks comprises three main aspects, which should be considered together wherever possible:

1   Evaluating the conditions of temperature and pressure to which the rocks were subjected during metamorphism

There is direct evidence from many laboratory experiments performed over the past 50 years that new minerals form at given temperatures and pressures. Rocks of different compositions are heated and squeezed in a laboratory under specified pressures and temperatures, and the resulting minerals are compared with those observed in naturally occurring metamorphic rocks. Pressures of up to 100 GPa can be obtained in the laboratory using a diamond-anvil cell (Figure 6.1). Note that a pressure of 1000 MPa (1 GPa) is equivalent to a depth of about 35 km beneath the Earth's surface.

The conditions of pressure and temperature that produced the naturally occurring metamorphic rocks are assumed to be the same as those imposed during the experiment. However, the assumption that rocks in the laboratory will behave similarly to rocks in the Earth's crust or mantle may not always be valid. Natural rocks are rather complex chemical systems. In addition to the major elements that make up the more abundant minerals, rocks contain a large number of minor elements that, although present in minute concentrations, can have a strong effect on the stability of metamorphic minerals. Equally important, minerals in natural rocks normally form in the presence of a fluid phase (usually a mixture of H_2O and CO_2), and the composition of this fluid influences which minerals are stable under the particular P–T conditions. Hydrothermal fluids of appropriate composition can be introduced into diamond-anvil cell experiments. Unfortunately, only rarely can the fluid phase that was present during the metamorphism of rocks be sampled, and the fluid compositions usually have to be inferred from the numerous minerals observed. There is also the problem of time. Natural rocks usually have millions of years in which to react to new conditions; but in the laboratory, the available timescale is restricted to days

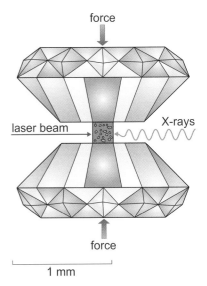

Figure 6.1   Diamond-anvil cells used for high-pressure experiments. A mineral sample is placed in a small hole (0.25 mm diameter) in metal foil inserted between two cut diamonds, each with 16 facets and weighing 0.125 carats. Diamonds are used because of their hardness and transparency to the X-rays needed to identify the resulting minerals. Temperatures are raised by laser beams impacting on the sample, which is inserted into a screw vice that squeezes the diamond tips together and imposes the required pressure on the sample.

or perhaps weeks. This is most important at low temperatures, because under such conditions the rates of reaction are very slow, and it is difficult to achieve equilibrium in the laboratory experiments. At higher temperatures, reactions proceed more rapidly, and so equilibrium can be achieved in a comparatively short space of time.

2   Thermal evolution and its relationship with the tectonic environment

In trying to interpret the significance of the *P–T* information gained from studying metamorphic minerals, the geologist must make sure it fits in with all the other geological evidence available. Heat for metamorphism comes from both internal sources (largely radiogenic heat), and from external sources such as the intrusion of hot magma. The precise combination of these sources determines the way that temperature increases with depth. This is recorded by the distribution of metamorphic minerals, and reflects the tectonic setting in which the rocks have formed. When metamorphic studies are combined with structural studies, the tectonic setting can often be identified, not only as a result of the highest metamorphic temperatures that the rocks have experienced, but also whilst the rocks were cooling during their ascent towards the surface of the Earth as a result of uplift and subsequent erosion. This process is called **exhumation**.

3   The relationship between metamorphism, as seen in the growth of new minerals, and deformation events recorded in large-scale and small-scale structures

A careful study of the mineral textures seen in thin sections and hand specimens provides the bridge between an estimate of metamorphic conditions and an understanding of the geological events that caused them: without it, we should learn little of the history of a metamorphic terrain. Such studies are based on textural evidence as discussed in Chapter 9.

In practice, the subject of metamorphism is limited to those transformations that take place between the surface zones of sedimentation and zones deep in the Earth's crust and mantle where partial melting begins. Such boundaries are gradational, and so inevitably there are areas of overlap in the study of the mineralogical and chemical changes that occur during the compaction and lithification of sediment, and the study of igneous petrology, where rocks start to melt. These two examples are of very low **metamorphic grade** and very high metamorphic grade, respectively.

The simplest definition of metamorphic grade is that higher grades of metamorphism reflect conditions of higher temperatures. It is a rather loose definition, but it will be defined more precisely later.

Activity 6.1   Metamorphic assemblages

In this activity, you will re-examine metamorphic rocks using the Virtual Microscope.

6.2   Metamorphic reactions

Rocks are solid objects and do not appear to react, at least at room temperature and atmospheric pressure. However, minerals, like most other chemical compounds, will react if temperatures and pressures are changed sufficiently. In fact, all mineral assemblages in metamorphic rocks result from chemical reactions that take place as the rock undergoes metamorphism. These reactions can be expressed in terms of either mineral names or chemical formulae.

For example, consider the reaction:

$$\text{muscovite} + \text{quartz} \rightleftharpoons \text{alkali feldspar} + \text{sillimanite} + H_2O \qquad (6.1)$$

H_2O will be a gas or a liquid depending on the P–T conditions of the reaction. The minerals muscovite and quartz are the solid **reactants** and alkali feldspar and **sillimanite** (a metamorphic mineral) are the mineral **products**. The same reaction can be expressed by an equation using the chemical formulae of the four minerals, and the fluid phase:

$$\underset{\text{muscovite}}{KAl_3Si_3O_{10}(OH)_2} + \underset{\text{quartz}}{SiO_2} \rightleftharpoons \underset{\substack{\text{alkali} \\ \text{feldspar}}}{KAlSi_3O_8} + \underset{\text{sillimanite}}{Al_2SiO_5} + H_2O \qquad (6.2)$$

There are two properties of chemical reactions we now need to introduce in order to understand metamorphic reactions better.

First, any chemical reaction involves a change in **entropy** (represented by the symbols ΔS, pronounced 'delta S') between reactants and products (i.e. the sum of the entropies of the products minus the sum of the entropies of the reactants). Entropy reflects the ordering of the atomic structure of matter. A system with zero entropy is perfectly ordered, whereas the same system with high entropy is more disordered. At any given pressure, a liquid or a gas is more disordered than a crystalline solid. Consider the chemical reaction involved in dissolving sugar in a cup of coffee. The reactants are sugar (a crystalline solid) and coffee (a liquid); the product is entirely liquid (sweet coffee), so the system has increased its atomic disorder and therefore increased its entropy.

■   To take another example, which has the lowest entropy: ice, water or steam?

☐   Ice, with a crystal structure, is the most highly ordered and so has the lowest entropy.

Second, any chemical reaction involves a change in **molar volume** as represented by the symbols ΔV ('delta V'). The molar volume is the volume occupied by one molecular weight of the mineral, gas or liquid. ΔV is defined as the change in volume between the sum of the molar volumes of the reactants and the sum of those of the products. A substance with a small molar volume has a closely packed atomic structure and is dense; conversely, a substance with a large molar volume is less dense. So a gas has a larger molar volume than a liquid, which in turn has a larger molar volume than a solid. When graphite is transformed into diamond at high pressures, there is a net increase in density and so a decrease in the molar volume.

■ Returning to Reaction 6.1, is the value of ΔS positive or negative? (You can assume that the four minerals involved have similar values of entropy.)

☐ H_2O is the only fluid (liquid or gas) involved. Fluids have less-ordered structures than any solid and so have higher entropies than any solid. Consequently, entropy of the products is greater than the entropy of the reactants. In other words, entropy increases as the reaction proceeds to the right and so ΔS is positive.

■ Is the value of ΔV positive or negative in Reaction 6.1?

☐ It is positive, because two solids have reacted to form two other solids and a fluid, which is generally much less dense (and so has a large molar volume) than any solid.

Entropy may also be used to derive a more satisfactory definition of metamorphic grade. Earlier (Section 6.1), we stated that higher grade is usually taken to mean higher temperature. We may now define metamorphic grade in terms of the entropy change (ΔS) of the metamorphic reactions concerned. *Increasing grade involves an increase in entropy of the metamorphic system* (including, of course, any gas phase).

■ In Reaction 6.1, which assemblage will be present at higher metamorphic grade: (i) muscovite and quartz, or (ii) sillimanite, alkali feldspar and H_2O?

☐ Assemblage (ii), sillimanite, alkali feldspar and H_2O, as this assemblage is less ordered, so has the higher entropy.

Most metamorphic reactions take place as the metamorphic grade increases, and are called **prograde reactions**; they tend to decrease the degree of ordering in the mineral system concerned.

Question 6.1

Another common metamorphic reaction is:

$$\text{tremolite} + \text{calcite} + \text{quartz} \rightleftharpoons \text{diopside} + H_2O + CO_2 \qquad (6.3)$$

where tremolite (a type of an amphibole) and diopside (a type of pyroxene) are both rich in Ca and Mg.

Determine: (a) which assemblage has the lower entropy, and (b) which assemblage would be stable at a higher grade of metamorphism.

For the muscovite + quartz reactions (Reactions 6.1 and 6.2), at a pressure of 150 MPa, temperatures of more than 550 °C are needed for muscovite and quartz to react to form alkali feldspar, sillimanite and H_2O. If alkali feldspar, sillimanite and H_2O are taken below 550 °C again, they will react together and revert to muscovite and quartz. How can it be then that alkali feldspar and sillimanite are often found in the same rock at room temperatures?

There are two reasons:

1   One of the products of the prograde reaction is a gas, H_2O, and the short answer is that it is not likely to stay around and wait for the rock to cool. Thus, if alkali feldspar and sillimanite are unaccompanied by H_2O at temperatures below 550 °C, they obviously cannot revert to muscovite and quartz.

2   The rates at which reactions take place vary enormously. In particular, they increase exponentially with increasing temperature, so that many reactions are extremely slow at low temperatures. Thus, if an assemblage is cooled rapidly, it may not have enough time to react and form low-temperature minerals.

In general, if conditions are such that reactions take place during cooling, they are called **retrograde reactions**; these occur as rocks readjust to the lower grades of metamorphism. The best textural evidence for retrograde metamorphism is provided by a relatively high-grade mineral surrounded, and partially replaced, by a mineral that is stable at lower metamorphic grades. For example, Figure 6.2 shows garnet partly being replaced by chlorite. The rock shows **textural disequilibrium** between the garnet and the chlorite because the reaction has not gone to completion. Retrograde reactions tend to occur during slow cooling in rocks that contain an H_2O-rich fluid phase. The reason that some garnet is preserved in Figure 6.2 is because either the reaction ran out of fluid, or the rock cooled too fast for the reaction to go to completion.

Figure 6.2   Garnet crystals (high relief) partly replaced by pale-green chlorite as a result of retrograde metamorphism (viewed in plane-polarised light; width of image = 1 mm).

Unfortunately, because the experimental problems are acute, very little is known about the actual rates of specific metamorphic reactions between coexisting solids. An exception is the change of aragonite to calcite. These two minerals are polymorphs: both have the same composition ($CaCO_3$) but different crystal structures, and the change from one to the other takes place in less than ten years at temperatures above 400 °C. The reaction has been studied experimentally at these temperatures and the results extrapolated to lower temperatures by plotting a graph of temperature against the time taken for all the aragonite to change to calcite (Figure 6.3). Note that such extrapolations can introduce comparatively large errors.

The experiments used to construct Figure 6.3 were done in the absence of water, but if water is present the times are greatly reduced. Even at 50 °C, all the aragonite would change to calcite in a mere million years, a consequence of which is that aragonite is very rarely found in fossils or as cement in ancient rocks.

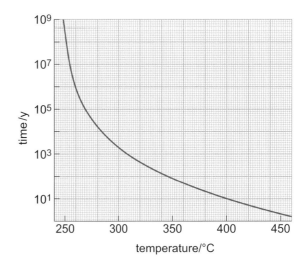

Figure 6.3   Time in years for complete change from aragonite to calcite as a function of temperature (anhydrous conditions).

Why should water speed up the change – bearing in mind that it cannot contribute chemically in a reaction in which one polymorph of $CaCO_3$ goes to a second polymorph of $CaCO_3$?

Water is an excellent **catalyst**, and is particularly important in speeding up metamorphic reactions that otherwise might not take place in the time available. In completely dry rocks, chemical changes can only occur by the slow process of ionic diffusion (the passage of ions through a solid crystal lattice). When water is present, it tends to spread itself along grain boundaries as an intergranular film, providing a network of chemical 'arteries' in which ions may move rapidly in solution and so speed up the metamorphic reactions.

This discussion of the muscovite + quartz reaction and the structural change from aragonite to calcite highlights three important general points:

1   Many prograde metamorphic reactions release H_2O and/or CO_2. Because these products are mobile and tend to move away from the rocks in which they were generated, they are often unavailable for retrograde reactions that might otherwise take place as the temperature falls.

2   Rates of reactions tend to be slow, but they increase exponentially with increasing temperature and are also speeded up by the presence of water.

3   Conversely, since many prograde metamorphic reactions release H_2O and CO_2, both tend to be driven off as metamorphism continues. Thus, by the time the highest grades are reached, almost all the fluid will have left the system, and recrystallisation and the growth of new minerals will have sealed off all the intergranular spaces, making the rock almost impermeable. Not only are H_2O and CO_2 not present to take part in reverse reactions, they are also unavailable as catalysts. Thus, such 'dry' rocks will then survive cooling more or less unaltered, because the rates of reaction are limited by the very slow rates of ionic diffusion. So high-grade mineral assemblages become 'frozen in', and preserved at low temperatures.

High-temperature minerals can be present at low temperatures, but they do not lie within their stability field when plotted on a phase diagram. Minerals that exist outside their stability fields are described as being **metastable**, which means that while they are not in a state of change, they can be made to change by some sort of impetus. For example, water can remain a liquid at temperatures below its freezing point, particularly if it is flowing. Freezing can be induced by halting its flow, or by adding a chunk of ice. For metamorphic assemblages, the impetus either comes from further metamorphism, when the rock is reheated and/or squeezed under a new set of P–T conditions, or from weathering, when the rock is attacked by atmospheric precipitation and its minerals are transformed into those that are in equilibrium at the temperatures and pressures present at the surface of the Earth – for example clay minerals in sediments.

■   How do we know that metamorphic rocks were in fact formed at high temperatures and pressures?

☐   From experiments designed to reproduce naturally occurring mineral assemblages under conditions of known pressure and temperature (Figure 6.1).

The results of such experiments are most simply illustrated using a phase diagram. In the experiments on which phase diagrams are based, there are three things that can be varied easily – the chemical composition of the sample, the temperature and the pressure. The minerals that form will depend on all three factors. It is obviously difficult to illustrate how they all vary on a two-dimensional diagram, so for convenience we usually keep one constant and show what happens as the other two vary. In most metamorphic experiments, it makes sense to work with a fixed composition, and to plot the results on a graph of pressure against temperature. Always remember, however, that such a *P–T* phase diagram only illustrates results for a particular composition; if the composition is altered, the details of the phase diagram will change.

Phase diagrams such as Figure 6.4a illustrate the stability ranges of three minerals all of which share the same composition, in this case Al_2SiO_5. These minerals are quite different in appearance. Kyanite forms blue blades (Figure 6.4b), sillimanite forms bundles of white fibres (Figure 6.4c) and andalusite forms prismatic needles with diamond-shaped cross-sections (Figure 6.4d).

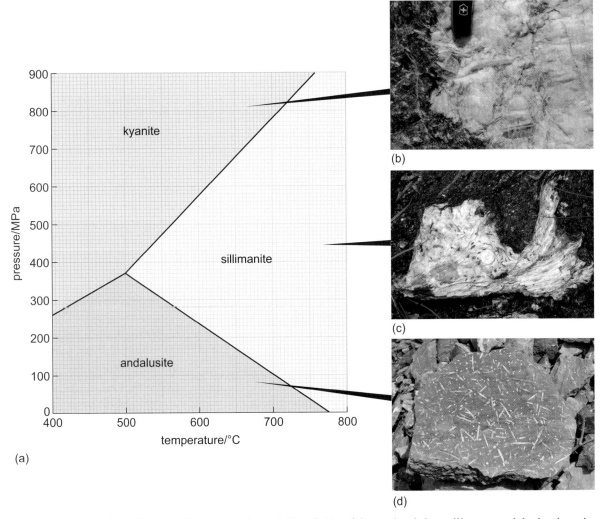

(a)

(b)

(c)

(d)

Figure 6.4   (a) Phase diagram illustrating the stability fields of three aluminium silicates: andalusite, kyanite and sillimanite. (b) Blades of blue kyanite in quartzite from the Himalaya (NW India). See also kyanite in the Digital Kit. (c) Sheaves of white fibres of sillimanite, locally discoloured by iron staining, from the Himalaya (NE India) (the coin is about 2 cm across). (d) White needles of andalusite in a spotted hornfels from the aureole of the Skiddaw granite, Cumbria, England. Andalusite is diamond-shaped in cross-section (width of image is 20 cm). See also andalusite in the Digital Kit.

Several points need to be emphasised in interpreting the phase diagram.

1   You may recall that a *phase* differs chemically and/or physically from the rest of the system being considered. In this case, kyanite, sillimanite and andalusite are three different phases because, although they share the same chemical composition, they have different crystal structures (i.e. they are polymorphs).

2   At equilibrium, the minerals kyanite, sillimanite and andalusite each exist over a range of different pressures and temperatures (Figure 6.4a). These are the conditions under which each mineral is stable, and the area depicting these conditions on a phase diagram is called its stability field. (Note also that to describe a mineral as 'stable' is synonymous with saying it is 'at equilibrium'.)

3   If a phase is present under conditions outside its stability field, it is said to be metastable.

4   The lines that mark the boundaries between the stability fields of kyanite, sillimanite and andalusite are phase boundaries. These lines mark the only places on this phase diagram where two of these minerals can exist together *at equilibrium.*

If either the pressure or temperature is changed so that conditions move off the phase boundary, only one of the minerals becomes stable. You have seen how such changes can take a very long time to complete (Figure 6.3).

Question 6.2

Use the Al_2SiO_5 phase diagram (Figure 6.4a) to answer the following questions.

(a) Is andalusite stable at 300 MPa and 700 °C?

(b) Which phase is stable at 800 MPa and 500 °C?

(c) At what temperature do kyanite and sillimanite coexist in equilibrium at a pressure of 700 MPa?

(d) Kyanite and andalusite were reported together in the same sample at 400 °C at atmospheric pressure (0.1 MPa). Which of these two phases is metastable (i.e. lies outside its stability field)?

Your answers to Question 6.2 demonstrate that in the case of the kyanite–andalusite–sillimanite system, it is comparatively easy to make very general statements about the conditions of pressure and temperature, depending on which mineral was present at equilibrium; however, it is difficult to be specific. This can be improved by using a sample in which two of the minerals occur together in equilibrium, since this indicates that conditions were somewhere along the line of the phase boundary. But can the estimates of pressure and temperature be made even more precise? Can we say precisely where along the phase boundary the minerals from a sample were in equilibrium? The short answer is no – unless help is available from elsewhere. What we need is some other mineral with a different stability field to be present in the same rock.

■ What other reaction have you looked at that also involves sillimanite?

☐ Reactions 6.1 and 6.2, in which muscovite and quartz react to form sillimanite, alkali feldspar and H_2O.

This reaction is plotted in Figure 6.5, and in this case, the phase boundary marks the boundary between the stability field of muscovite and quartz and that of Al_2SiO_5, alkali feldspar and H_2O. Note that the term sillimanite has been replaced with the formula Al_2SiO_5 since other polymorphs of this composition are stable within the P–T conditions shown. The boundary defines the conditions at which the five phases can coexist together in equilibrium. Because the gradient of this slope is fairly steep, we can say that, for normal crustal pressures ($P < 1000$ MPa), all five phases will coexist over a temperature range of about 400–800 °C. But how can a more precise estimate be obtained?

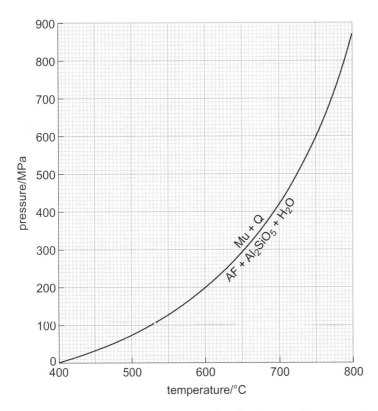

Figure 6.5 Pressure–temperature plot for the reaction muscovite (Mu) + quartz (Q) ⇌ alkali feldspar (AF) + Al_2SiO_5 + H_2O.

Figure 6.6 combines the phase diagrams for sillimanite–kyanite–andalusite and for muscovite–quartz–Al_2SiO_5–alkali feldspar–H_2O. By using all four phase boundaries, the diagram becomes subdivided into smaller areas of pressure and temperature (A, B, C, D and E) in which particular combinations of minerals are stable. You can see, for example, that muscovite and quartz only coexist in equilibrium with andalusite in area A.

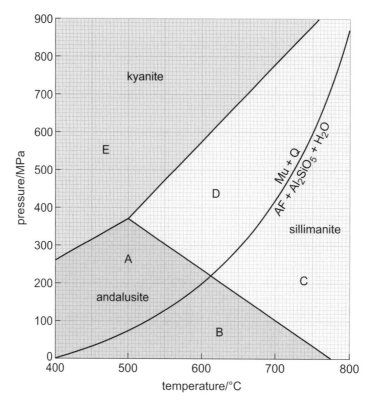

Figure 6.6   Phase diagram combining information from Figures 6.4a and 6.5.

There is one point on Figure 6.6 where six phases can occur together. So, if muscovite, quartz, alkali feldspar, andalusite, sillimanite and H_2O are found in equilibrium in one sample, it can only have crystallised at a particular pressure and temperature.

■   What would that pressure and temperature be?

☐   These six phases only coexist at about 610 °C and 220 MPa (where fields A, B, C and D meet at a point on Figure 6.6).

In summary, it should be apparent that:

1   It is not possible to obtain precise estimates of pressure and temperature on every sample of metamorphic rock. Usually, all you can say is that the rock crystallised somewhere in the stability fields of its particular minerals.

2   To obtain a reasonable estimate of metamorphic conditions, it is necessary to seek out rock specimens that contain certain key metamorphic minerals, and so may reflect conditions on a particular phase boundary. In some areas, suitable samples may be scarce or non-existent.

In general, it is necessary to combine information from different metamorphic reactions in an attempt to limit the possible range of pressure and temperature in which the rock could have crystallised. The grid-like pattern that is formed as more and more boundaries are added to the phase diagram is called a **petrogenetic grid**, and Figure 6.6 is a simple example.

There is another way of determining pressure and temperature of which you should be aware. In contrast to the minerals discussed so far in this section, many minerals have a range of compositions – the plagioclase feldspars, for example, range in composition between albite and anorthite by the interchange of the elements Na and Si with the elements Ca and Al. Reactions involving such minerals will not occur along a line on a phase diagram but be spread out across a range of P–T conditions.

For example, there is a reaction involving the minerals garnet, **cordierite**, quartz and any one of the three aluminosilicates (Al_2SiO_5):

$$\text{garnet} + \text{quartz} + Al_2SiO_5 \rightleftharpoons \text{cordierite} \tag{6.4}$$

Cordierite (Figure 6.7a) is a silicate mineral found in some aluminous metamorphic rocks (Figure 6.7b). Its composition varies from $Mg_2Al_4Si_5O_{18}$ to $Fe_2Al_4Si_5O_{18}$. Similarly, garnet in such rocks has a composition that varies from $Mg_3Al_2(SiO_4)_3$ to $Fe_3Al_2(SiO_4)_3$. The position of the phase boundary of the reaction described by Reaction 6.4 depends on the precise Fe/Mg ratio of the rock. In Figure 6.8, this phase boundary is plotted for both the iron-rich compositions and the magnesium-rich compositions, determined from experiments containing only Fe and only Mg compositions, respectively. For naturally occurring rocks containing both Fe and Mg, the reaction will lie somewhere between these boundaries, depending on the Fe/Mg ratio of the rock. For example, Figure 6.8 also shows the phase boundary for a natural cordierite with Fe/Mg = 1.5 (a typical value). For reactions involving such minerals, the geologist must not only identify which minerals are present but also analyse the precise composition of the cordierite and garnet to find the appropriate position of the phase boundary at which the two minerals coexist in equilibrium with an aluminosilicate and quartz.

(a)

(b)

Figure 6.7 (a) Cordierite (blue mass of crystals in quartz-rich patch about 2 cm across) and biotite (black) crystals from (b) migmatite outcrop, central Madagascar (lens cap for scale).

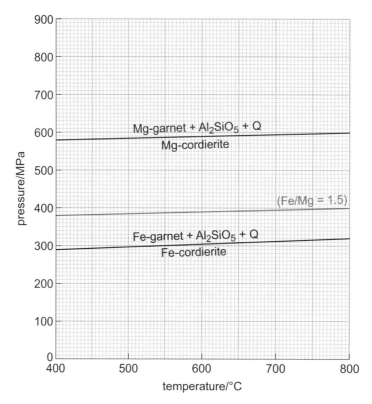

Figure 6.8   Phase boundaries for Reaction 6.4 using Fe and Mg end members. Also shown is the phase boundary for a cordierite with the Fe/Mg ratio = 1.5 (the middle (red) line).

■ What is the main difference between the phase boundaries shown on Figure 6.8 and that on Figure 6.5?

☐ The phase boundary in Figure 6.5 is much steeper.

■ Which reaction gives a better idea of the pressure at which the phases on either side of the boundary can coexist?

☐ The reaction involving cordierite, garnet, sillimanite and quartz, as it gives a narrow range of pressures for a given Fe/Mg ratio.

■ If a rock contains negligible Mg, what is the pressure range over which the mineral cordierite, garnet, sillimanite and quartz can coexist?

☐ From the lower phase boundary in Figure 6.8, a range of 290–320 MPa is indicated.

The reason that this reaction is nearly parallel to the temperature axis is that it involves a large change in molar volume (ΔV) but only a small change in entropy (ΔS).

The large change in volume is because garnet is a denser phase (density = 3.58×10^3–4.32×10^3 kg m^{-3}) than cordierite (2.53×10^3–2.78×10^3 kg m^{-3}). Reactions with a gentle slope, such as Reaction 6.4, make good natural barometers and those with a steep slope, such as Reaction 6.1, make good natural thermometers of metamorphic conditions. Good natural barometers are provided by reactions with a large change in molar volume (ΔV) (but should not also involve a large change in entropy), and good natural thermometers are provided by reactions with a large change in entropy (ΔS).

So far we have stressed the importance of pressure and temperature in determining which minerals are likely to be stable.

■ What is the third, and arguably most important, factor that governs whether a particular metamorphic mineral is stable?

☐ The bulk chemical *composition* of the rock being metamorphosed: rocks of different composition tend to contain different minerals, even if they have been at equilibrium under the same conditions of temperature and pressure (Book 1, Section 8.3).

The reason for the importance of chemical composition is simple. If a rock does not contain the appropriate elements for a particular mineral, then that mineral cannot be formed.

Question 6.3

Which of the minerals garnet, calcite, muscovite and feldspar are likely to be present in the metamorphosed equivalents of: (a) basalt; (b) limestone; and (c) mudstone? Explain your answer in each case.

The important point to remember is that if metamorphic rocks contain different minerals but have been exposed to the same temperatures and pressures, then they must have different chemical compositions.

6.3   Types of metamorphism

At the beginning of this chapter, we described metamorphic processes as those that are associated with mineralogical and structural changes in rocks. You have seen a few of the sorts of reactions that take place, the factors that control them, and the methods used to estimate the conditions of temperature and pressure. You will now look a little more closely at the kinds of rocks that metamorphism produces, and the three main types of metamorphism.

6.3.1   Dynamic metamorphism

Dynamic metamorphism takes place in areas of intense *local* deformation, such as in fault zones or **shear zones**. Temperatures and pressures may be low if dynamic metamorphism occurs at high levels in the Earth's crust where rocks are brittle, but at deeper levels, higher temperatures result in more 'ductile' structures in which flow predominates over fracture. The differences between brittle and ductile deformation will be examined more fully in Chapter 7.

In a major fault zone, there is mechanical movement of one rock over another so that any rocks within the fault zone may be ground down in a process known as **cataclasis** (from the Greek meaning 'break down'). The resulting rocks are called cataclastic rocks. Almost all brittle faults contain a zone in which there are cataclastically broken or crushed rocks known as a **fault breccia** (Figure 6.9).

Figure 6.9   Fault breccia from Barra, Outer Hebrides.

More ductile deformation causes slippage between layers within the rocks, and between planes of atoms within minerals. Recrystallisation under these conditions results in a fine-grained foliated rock with a streaked-out texture indicating the direction of shearing, which is known as a **mylonite** (Figure 6.10a). During the formation of mylonites, quartz is broken down into fine-grained aggregates that form elongated ribbons (Figure 6.10b). During mylonite formation, or indeed any other kind of dynamic metamorphism, the grain size is reduced.

(a)

(b)

Figure 6.10   (a) Mylonite from the Moine Thrust, Scotland. (b) Ribbon texture from recrystallised quartz in a mylonite (viewed between crossed polars; width of image = 5.5 mm).

Movements along fault planes expend considerable amounts of energy to overcome friction and most of this energy is released as *heat*. Although the size of the resultant increase in temperature depends on (i) frictional heating along the fault plane, and (ii) the thermal conductivities of rocks on either side, it appears to be most sensitive to (iii) the rate of movement of the fault. Earthquakes are produced by rapid movements on faults and it can be shown that movements of this kind, although of very short duration, can produce a temperature rise sufficient to cause melting in the deformed rock – about 5 cm per second has been suggested as the sort of speed at which rock melting might be expected.

Rocks melted by frictional heat are rare, but they do turn up in small quantities all over the world. Such melts form black, fine-grained rocks that look like basaltic glass and because they intrude the country rocks adjacent to the fault zone in small irregular dykes and veins (Figure 6.11), they have often been identified wrongly as igneous rocks of more conventional origin, hence their name **pseudotachylites** (tachylite being an old term for basaltic glass).

6.3.2   Contact metamorphism

Contact metamorphism (also known as thermal metamorphism) refers simply to the metamorphic changes taking place in response to the heat associated with igneous bodies. It is most obvious around intrusive rocks because, unlike their extrusive equivalents, most of their heat is not lost to the atmosphere but is dissipated into the surrounding country rocks. Naturally, the temperatures are highest close to the igneous body itself, so there is therefore a very marked increase in metamorphic grade near the contact. The zone of metamorphic rocks around the intrusion is termed a metamorphic aureole.

Figure 6.11   Pseudotachylite (cross-cutting dark band) within a biotite migmatite gneiss from Uist, Outer Hebrides.

In a metamorphic aureole, high-temperature minerals will grow nearest to the contact and low-temperature minerals are found furthest away. We can plot on a map where these metamorphic minerals occur in just the same way as we can record any other features, and in so doing would map **metamorphic zones**. Each zone is known by the name of a particular characteristic mineral (termed an **index mineral**) present in that zone; for example, the biotite zone or the garnet zone. Provided rocks of similar composition are being examined, the index mineral is a qualitative indication of the metamorphic grade. This approach is extremely useful where rocks of similar compositions are exposed over a large enough area, because the metamorphic zones provide an indication of the relative metamorphic grades, even when precise estimates of pressure and temperature are not available.

For example, the aluminosilicate polymorph that occurs within metamorphosed mudstones very close to their contact with the granites of southwest England is andalusite (Figure 6.4d). The distribution of andalusite around a granite defines the boundaries of the *andalusite zone*. Farther away from the granite, where the temperature of the rocks reached during or after intrusion was lower, andalusite is not found.

In this way, a series of metamorphic zones can be mapped out around an intrusion, each zone representing successively lower temperatures away from the body as, for example, has been done around the Ardara granodiorite in Donegal

(Figure 6.12). In this example, the three aluminosilicates kyanite, andalusite and sillimanite are particularly useful.

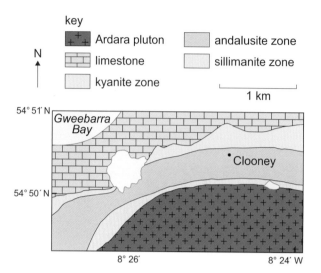

Figure 6.12 Metamorphic aureole around the Ardara pluton, Donegal, Ireland.

■ With reference to Figure 6.4a, determine whether, for any particular pressure, sillimanite is stable at higher or lower temperatures than andalusite. Is that consistent with their relative positions in the contact aureole in Figure 6.12?

☐ Sillimanite is always stable at higher temperatures than andalusite at a fixed pressure (Figure 6.4a), which explains why it occurs nearer to the granite contact than andalusite does in Figure 6.12.

Question 6.4

(a) Since andalusite rather than kyanite reacts to form sillimanite in the contact aureole of the Ardara pluton, what is the maximum pressure at which metamorphism could have taken place?

(b) The pressure at which the granodiorite was emplaced was 200 MPa. Use the phase diagram (Figure 6.4a) to estimate the temperature at the outer margin of the sillimanite zone.

(c) Estimate the depth at which the intrusion was emplaced.

In general, large-scale changes in the bulk chemical composition of the original rocks do not occur during metamorphism. This applies to contact metamorphism, but with some important exceptions. Right up against the igneous contact, elements can be introduced from the crystallising magma. For example, carbonate-rich rocks are sometimes replaced by calcium and magnesium silicates, iron oxides and sulfides.

■ What is the name of the process that changes the bulk composition of rocks during the influx of hydrothermal fluids?

☐ Metasomatism (see Section 4.2).

In some metamorphic aureoles, metasomatism has resulted in the concentration of economic amounts of ores. One such example occurs around the Skiddaw granite in Cumbria, northwest England. Extensive hydrothermal activity has resulted in the alteration of the granite near its margins and the development of mineralised veins that have been exploited by tungsten mining. The granite itself forms a flat-topped intrusion of biotite granite of Devonian age (400 Ma) intruded into Ordovician slates and siltstones (Skiddaw Group, Figure 6.13). An elliptical aureole has developed within the sediments around the granite. Hard hornfels is developed near to the granite, with spotted slates developed in the outer parts of the aureole. The minerals formed in these zones carry important information about the conditions at which the granite has formed.

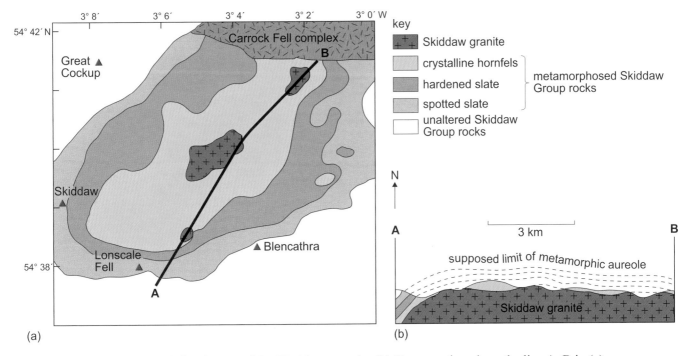

Figure 6.13 (a) Geological sketch map of the Skiddaw aureole. (b) Cross-section along the line A B in (a).

Finally, the textures of rocks in a thermal aureole can be distinctive. Since the emplacement of magma may not cause a preferred alignment of recrystallised minerals (at least at shallow crustal levels), the main characteristic of rocks that have undergone contact metamorphism is the formation of a hard, compact and splintery rock called a hornfels. Frequently, the first sign of contact metamorphism in the metamorphic aureole is a distinctive 'spotting' caused by the growth of clusters of new metamorphic minerals. Such a rock is called spotted hornfels, as shown in Figure 6.4d.

Activity 6.2 The Skiddaw granite and its metamorphic aureole

This activity brings together the field relations, thin sections and pressure–temperature diagram from a metamorphic aureole in Cumbria, using the Digital Kit.

6.3.3   Regional metamorphism

Regional metamorphism is a very much larger-scale phenomenon than either dynamic or contact metamorphism. Regional metamorphism is usually associated with continental collision or subduction, and therefore with deformation. At the highest grades of metamorphism, rocks begin to melt and ultimately granites are formed.

The patterns of changing metamorphic grade across an area differ between contact and regional metamorphism. In contact metamorphism, the zones tend to be concentric around a particular intrusion, whereas in areas of regional metamorphism, the zones are often linear – as you might expect if they represent deeply eroded mountain belts, formed along some ancient convergent plate boundary or collision zone.

The third and perhaps most important distinction between contact and regional metamorphism is that the sources of heat are different, which has important implications for the rate at which metamorphism occurs. During contact metamorphism, the heat source is a cooling igneous body. This is an effective heat source only for as long as the magma is crystallising. Even for a large granite body, this will be less than 100 000 years. For a dyke, heat may be exhausted after 100 years or less. During regional metamorphism, the heat source comes largely from radiogenic heat from the decay of heat-producing elements. This is a much slower process. For example, garnets from metamorphosed sedimentary rock in the Himalaya have grown during regional metamorphism following continental collision (Figure 6.2). Highly precise dating of the mineral garnet has shown that the core of a crystal is 30 Ma old, and that successively younger overgrowths were formed towards the outer parts of the garnet which are only 25 Ma old. This study indicates that during regional metamorphism rocks can be heated for at least five million years.

Our present understanding of the metamorphism at continental collision zones is possible because of the foundations laid by a handful of 19th century geologists. In 1893, George Barrow made a classic study of the regionally metamorphosed rocks of the Scottish Highlands. He discovered a 10–20 km thick sequence of sedimentary and volcanic rocks that had been highly deformed and metamorphosed about 470 million years ago during what is now known to have been a period of subduction and collision. Barrow mapped these rocks and studied their thin sections. By considering the sequence of metamorphosed mudstones and shales, often called **pelitic rocks** (or pelites), he was able to establish a series of zones, each characterised by a distinctive mineral (Figure 6.14).

Barrow was the first person to recognise metamorphic zones and to use index minerals to identify successive grades of metamorphism from the appearance of distinctive new minerals in a traverse across rocks of successively higher grade. Each new

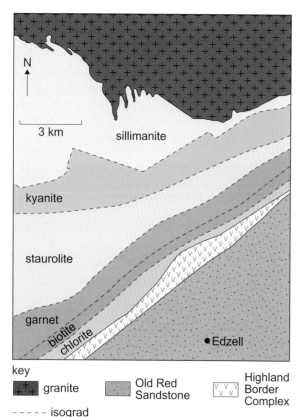

key

- granite
- Old Red Sandstone
- Highland Border Complex
- - - - - isograd

Figure 6.14   Geological sketch-map of the Barrow zones of Glen Esk (Angus), southeast Highlands of Scotland. Note that the Highland Border Complex and the Old Red Sandstone are part of younger sequences, largely unaffected by the regional metamorphism.

mineral represents a further step towards higher temperatures and/or pressures, and so becomes an index mineral indicative of metamorphic grade. He recognised six metamorphic zones in the metamorphic rocks of the southeast Highlands. The index minerals he based them on were (in order of increasing metamorphic grade) chlorite, biotite, garnet, staurolite (a brown iron–aluminium silicate, Figure 6.15), kyanite and sillimanite; of these minerals, kyanite is the most spectacular (Figure 6.4b).

What can we say about the conditions of pressure and temperature reflected in the appearance of these index minerals? Once again, the approach is very simple in principle – experiments are carried out at known pressures and temperatures and the reactions that take place are then observed. The change in *P–T* conditions between the kyanite and sillimanite zones, for example, can be interpreted by considering the experimentally determined phase diagram in Figure 6.4a.

■   How do the aluminosilicates recognised by Barrow in the southeast Highlands compare with those present in the metamorphic aureole in Figure 6.12?

☐   Andalusite is absent from the Barrow zones.

■   Do the minerals in the Scottish examples reflect higher or lower pressures than those in the metamorphic aureole?

☐   In Figure 6.4a, the phase boundary between kyanite and sillimanite exists only at pressures greater than 370 MPa, whereas the phase boundary between andalusite and sillimanite is at less than 370 MPa. Thus, the transition from kyanite directly to sillimanite in the Scottish rocks indicates that they were formed at higher pressures (>370 MPa) and therefore greater depths.

Question 6.5

Figure 6.16 is the pressure–temperature diagram for a reaction in which garnet is formed. Having argued that the pressure in this area was greater than 370 MPa, use Figure 6.16 to infer the approximate temperature at which garnet first appears, that is, at the beginning of the garnet zone.

In the case of contact metamorphism, the near-concentric mineral zones around an igneous body essentially represent zones of successively higher temperatures. Although the same principles apply, regional metamorphism is more complicated, since both pressure and temperature vary. In both types of metamorphism, however, the points of first appearance of an index mineral can be mapped and linked by a line. Provided such minerals are *in rocks of similar composition*, this line marks the position of rocks with 'equal metamorphic grade' and so is known as an **isograd**. Conventionally, an isograd is named after the index mineral on the side of *higher* metamorphic grade. Thus, the border between the kyanite and sillimanite zones in Figure 6.14 is marked by the appearance of sillimanite as an index mineral and is called the sillimanite isograd.

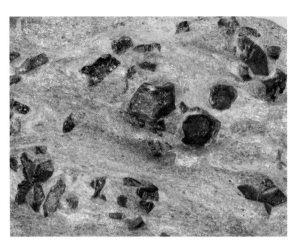

Figure 6.15   Muscovite schist with large red garnet crystals (centre of image) and several dark staurolite porphyroblasts from Karakoram, Pakistan (field of view 35 mm across). Note twinned staurolite in lower left of image – staurolite takes its name from such cross-shaped twins.

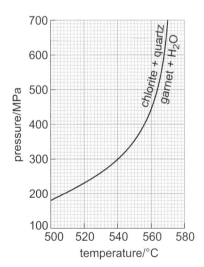

Figure 6.16   Pressure–temperature diagram for the reaction chlorite + quartz ⇌ garnet + H_2O.

It could be argued from Figure 6.4a that the transition from kyanite to sillimanite does not *necessarily* involve an increase in temperature, but could reflect a decrease in pressure. This possibility can be excluded by a more detailed examination of the minerals on either side of the isograd – and, in general, sequences of metamorphic zones are determined by changes of temperature rather than by changes of pressure.

Question 6.6

Figure 6.17 is a geological sketch map illustrating some folded metamorphosed sediments (metasediments) and a gabbro intrusion.

(a)   Sketch in and label the isograds between the different mineral zones.

(b)   Is the metamorphism older or younger than the folding?

(c)   Does the metamorphism appear to be related to the gabbro intrusion (i.e. is it an example of contact metamorphism)?

The isograds you drew on Figure 6.17 may look convincing, but do not assume that they can be drawn as precisely as, for example, a topographic contour line. Rocks have complex chemical compositions involving many elements and thus represent messy chemical systems. None the less, several experienced geologists working in the same area would usually put an isograd at roughly the same place on the map.

The significance of isograds for the rate of temperature increase with depth, or **geothermal gradient**, during metamorphism may not be clear cut. In going from south to north across Barrow's zones, you are going from an area of low-grade metamorphism to one where both the temperatures and pressures were higher. Whether the two areas record metamorphism at the same time is a complex question beyond the scope of this course.

Finally in this section, we must reiterate the importance of rock composition in determining which metamorphic minerals will crystallise. In a sequence of pelites, calcic and mafic rocks, different minerals will grow in each rock type and so there will be different index minerals in each rock type. You have seen that garnet, staurolite, kyanite and sillimanite can be used as important index minerals to establish metamorphic zones; they are all rich in aluminium, and they occur in Al-rich metamorphosed sediments (pelites). Fewer mineralogical changes occur in carbonates and mafic igneous rocks and so pelites are generally more useful in establishing metamorphic zones. However, it is important to realise that if, as is common, metamorphic zones are based on index minerals found in pelites, the zones may well include rocks of other compositions. So there is nothing wrong with talking about mafic igneous rocks being in the staurolite zone, even though they contain no staurolite.

Question 6.7

Figure 6.18 illustrates a series of isograds based on minerals in pelitic rocks. The isograds cut across a series of calcic, pelitic and mafic igneous rocks. Use Table 6.1 to indicate what mineral assemblages you would expect to find at points A, B, C and D.

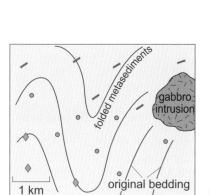

Figure 6.17   Geological sketch map for use with Question 6.6. Different symbols identify the index mineral found at various exposures.

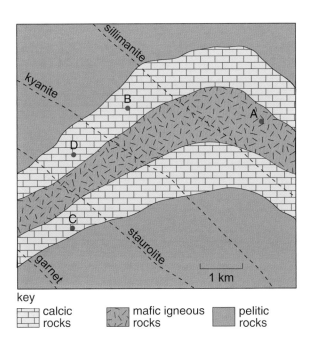

Figure 6.18   Isograds (based on pelitic rock minerals) cross-cutting a sequence of calcic, pelitic and mafic igneous rock types. (For use with Question 6.7.)

key

calcic rocks

mafic igneous rocks

pelitic rocks

Table 6.1   Metamorphic assemblages produced in pelitic, calcic (originally carbonate-rich) and mafic igneous rocks.

| Metamorphic zone (based on pelites) | Assemblage produced (index mineral in italics) | | |
	Pelites	Calcic rocks (calcite may also be present)	Mafic igneous rocks
garnet	*garnet*, mica, quartz, Na-rich plagioclase	garnet, epidote, amphibole	chlorite, albite, epidote
staurolite	*staurolite*, garnet, quartz, mica, Na-rich plagioclase	garnet, anorthite, amphibole	amphibole, plagioclase
kyanite	*kyanite*, garnet, quartz, mica, Na-rich plagioclase	garnet, anorthite, amphibole	amphibole, plagioclase
sillimanite	*sillimanite*, garnet, quartz, mica, Na-rich plagioclase	garnet, pyroxene, anorthite	amphibole, plagioclase, garnet

6.4   Metamorphic facies

The mapping of metamorphic zones using index minerals is very useful for describing metamorphism from a particular area. However, it is difficult to compare the metamorphic grades of different areas on the basis of index minerals because of the possibility of different bulk compositions in the two areas, which will result in different metamorphic index minerals even under similar conditions of pressure and temperature. **Metamorphic facies** is a term coined by the Finnish geologist Pentti Eskola (1883–1964) to embrace *all* the possible metamorphic mineral assemblages produced in rocks of different composition at similar temperatures and pressures. The assignment of a rock to a metamorphic facies is based on the observed mineral assemblage, which corresponds to a particular P–T range of metamorphism.

This definition requires amplification:

1   The composition of a metamorphic rock determines its mineral assemblage under any particular conditions of temperature, pressure and fluid composition. Thus, given a chemical analysis of a rock, it should be possible to predict the mineral assemblage that is likely to be stable under a given set of physical conditions (Table 6.1).

2   A metamorphic facies is determined from the *observed mineral assemblages* in associated rocks: it does *not* require knowledge of the precise conditions of temperature and pressure, which are estimated subsequently from a petrogenetic grid.

The first point you have met before, but the second is important because it often causes confusion. You have seen how metamorphic conditions may be estimated from experimental results on particular metamorphic reactions. Moreover, when data on a number of reactions that occur under different conditions are plotted together on a *P–T* diagram, a petrogenetic grid is formed, which is a criss-cross of different phase boundaries that allows likely metamorphic conditions to be inferred from particular mineral assemblages. But such a petrogenetic grid is obviously experimentally determined, and thus, although its general form is unlikely to change significantly, it will be continually modified as experimental techniques develop and more experiments are carried out, particularly in the presence of fluid phases of different compositions. A metamorphic facies, by contrast, is assigned to the rock from the observed mineral assemblages. Provisional limits of temperature and pressure may then be assigned to the particular metamorphic facies from the petrogenetic grid as currently understood.

Let us take an example that you have already met. In Barrow's area of the Scottish Highlands, pelitic rocks in the higher grades of metamorphism contain sillimanite, whereas the calcic rocks contain garnet and pyroxene. Although the minerals found in the two types of rock are different because the rocks have different chemical compositions, both assemblages belong to the same metamorphic facies. In areas of lower-grade metamorphism, pelites contain chlorite and biotite, and the calcic rocks contain calcite, epidote (a Ca-rich chain silicate) and an amphibole, both rock types belonging to the same metamorphic facies.

■   What metamorphic zone would the mineral garnet indicate if found in a calcic rock?

☐   From Table 6.1, garnet is present in all metamorphic zones in calcic rocks. It does not therefore help to define the particular zone.

■   Will a garnet-bearing rock of pelitic composition always be from the same metamorphic facies as a garnet-bearing rock of a calcic composition?

☐   No. Because they can be from different metamorphic zones, they can also be formed at different pressures and temperatures. If so, the metamorphic facies may be different as well.

Today, geologists recognise eight principal metamorphic facies that cover temperatures from 100 °C to over 800 °C, and pressures from atmospheric pressure to over 1400 MPa (Table 6.2). One of these (eclogite) you have met before in

Section 5.4.4. There, it was defined as a particular rock type; a metamorphic rock of mafic composition with garnet and pyroxene present (Figure 6.19). Here, we are using the term **eclogite facies** to indicate the *P–T* field over which such an assemblage is stable (i.e. a metamorphic facies). Although it is not necessary to define each facies here, the facies names generally result from the stable assemblage found in rocks of mafic composition. For example, **blueschist facies** rocks contain blue minerals, such as the amphibole, glaucophane, only if the bulk composition is mafic (Figure 6.20). A pelite, when metamorphosed under blueschist-facies conditions, will be made up largely of white mica and quartz, but neither of these is blue! None the less, if we know that it formed under these conditions it would be correct to say it is a blueschist-facies rock.

Table 6.2 Summary of metamorphic facies and their likely environments of formation. Zeolite, prehnite and pumpellyite are all low-temperature calcic aluminosilicates.

Facies	Environment
zeolite	buried sediments
prehnite–pumpellyite	more deeply buried sediments
blueschist	relatively high pressure but very low temperature: subduction zones
greenschist	extremely widespread facies, low-grade regional metamorphism in orogenic areas
amphibolite	widespread facies, high-grade regional metamorphism in orogenic areas
granulite	highest grade of regional metamorphism, occurs in the lower crust and within some Precambrian cratons
hornfels	metamorphic aureoles: moderate to high temperatures but at low pressures
eclogite	very high pressures and moderate to high temperatures

(a)

(b)

Figure 6.20 (a) A blueschist-facies locality from western Turkey. The blue tinge results from the presence of glaucophane. (b) Polished block (width = 5 cm) of blueschist with glaucophane porphyroblasts; also present are muscovite, garnet and epidote. Glaucophane can be observed using the Virtual Microscope by viewing the blue amphibole present in the metamorphic rock specimen 'Amphibolite'.

Figure 6.19 Eclogite, a spectacular, high-density rock from Norway. The principal minerals are garnet (pink) and a sodium-rich pyroxene (green) called jadeite (width of image = 150 mm).

Our present estimates of the *P–T* fields that characterise the eight metamorphic facies are reproduced in Figure 6.21. Broadly, metamorphic facies may be linked to estimates of pressure and temperature – and so to the change in temperature with depth known as the geothermal gradient. For example, hornfels results from contact metamorphism, which is characterised by high but localised geothermal gradients due to the associated magmatic activity. Very low geothermal gradients occur when cold material from the Earth's surface is taken down to lower crust or upper mantle depths. In between, there are more 'normal' geothermal gradients, which characterise much of the Earth's continental crust.

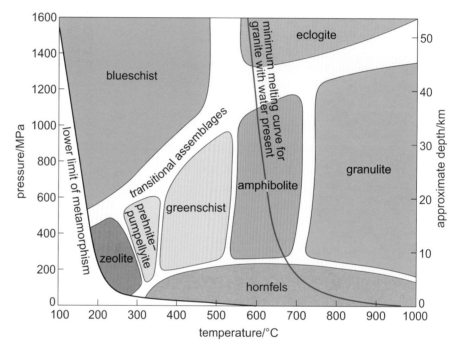

Figure 6.21   The *P–T* fields of the major metamorphic facies.

◼   If you plot a typical continental geothermal gradient of 30 °C km^{-1} on Figure 6.21, what sequence of metamorphic facies would you expect to find with increasing depth in the Earth's crust? Assume a surface temperature of 0 °C.

☐   The geotherm starts at 0 °C at the surface and increases by 300 °C for every 10 km of depth. Near the surface, it passes through **zeolite** and **prehnite–pumpellyite facies**, then successively **greenschist** and **amphibolite facies**, reaching **granulite facies** at a depth of about 25 km.

This observation raises two important issues. First, the occurrence of rocks indicative of metamorphic facies not in this sequence suggests that the Earth's geothermal gradient can be significantly disturbed. Second, how and why are rocks formed in higher pressure metamorphic facies subsequently brought back up to the surface? One explanation is that their exhumation results from isostasy in response to unusual thicknesses of continental crust. As an example, take Barrow's sillimanite isograd, which was formed at a pressure of 680 MPa, equivalent to a depth of ~25 km during metamorphism. At present, these rocks are exposed on top of a 35 km-thick crust. The simplest interpretation is that during the metamorphism that piece of crust was at least 25 + 35 km, or 60 km, thick – compared to an average thickness for continental crust of 30–35 km. Since then, the thick crust has been uplifted, the top 25 km has been eroded, and the sillimanite isograd exhumed and exposed at the surface.

■ The metamorphic facies diagram covers a very wide range of temperatures and pressures. Other than metamorphism, what other effects might occur at very high temperatures?

☐ Clearly, if rocks are heated to a high enough temperature, they are going to melt. On Figure 6.21, the red line represents the melting curve for rocks of granitic composition when water is present (it is the water-saturated solidus, see Figure 5.3). At pressures greater than 400 MPa, melting takes place at about 600 °C, in the presence of H_2O.

Melts formed during regional metamorphism may rise to shallower levels and even cause their own metamorphic aureole. Thus, the effects of thermal and regional metamorphism may cause complex changes in metamorphic grade both horizontally and vertically.

6.5   Plate tectonics and metamorphism

The sequence of metamorphic facies recorded during deformation of the Earth's crust reflects the complex interaction between available heat sources and structural evolution. Some associations of metamorphic facies are indicative of particular tectonic settings and we shall examine two of these in a little detail.

6.5.1   Metamorphism and subduction

In general, the dominant heat source for regional metamorphism is supplied internally by the decay of heat-producing elements (Section 2.4). The geothermal gradient $\dfrac{\Delta T}{d}$ varies from one tectonic environment to another, but is related to the **heat flow** (q) by the equation:

$$q = \frac{K \Delta T}{d} \tag{6.5}$$

where K is the **thermal conductivity** of the rocks and ΔT is the change in temperature that occurs over a depth range d. Heat flow is a measure of the amount of energy passing through unit surface area in unit time, usually expressed in milliwatts per square metre, mW m^{-2}.

Over much of the Earth's surface, the heat flow is surprisingly uniform at about 60 mW m^{-2}. Striking exceptions occur near mid-ocean ridges and island arcs where the heat flow is high, and above oceanic trenches at convergent plate boundaries where it is low. This is illustrated in a profile of the heat flow observed across the western Pacific in Figure 6.22.

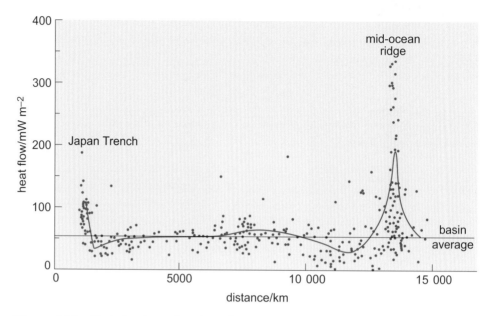

Figure 6.22   Variation in surface heat flow, measured across the western Pacific Ocean. The curved red line represents the average local heat flow, and the horizontal line illustrates the average heat flow from all ocean basins.

The low heat flow values on the flanks of the mid-ocean ridge in Figure 6.22 are due to cooling caused by the circulation of considerable volumes of seawater through the top few kilometres of the newly formed ocean crust.

■ Why are the heat flow values so much higher in Japan and on the mid-ocean ridge?

☐ These are both areas where considerable volumes of magma are intruded. Magmas are hot, and as they move up towards the surface they bring heat with them, thereby increasing the heat flow.

■ Assuming that the thermal conductivity does not change significantly, will the geothermal gradient be higher or lower in areas of higher heat flow?

☐ Higher, since the geothermal gradient is proportional to the heat flow (Equation 6.5).

■ What might be the cause of the low heat flow values in the Japan Trench?

☐ Such trenches mark the site of active subduction (see Figure 2.4). The ocean crust has been at or near the surface. It is therefore relatively cold, and as oceanic lithosphere is subducted into the mantle the isotherms are also dragged down resulting in a low geothermal gradient and thus low heat flow to the surface (Figure 6.23).

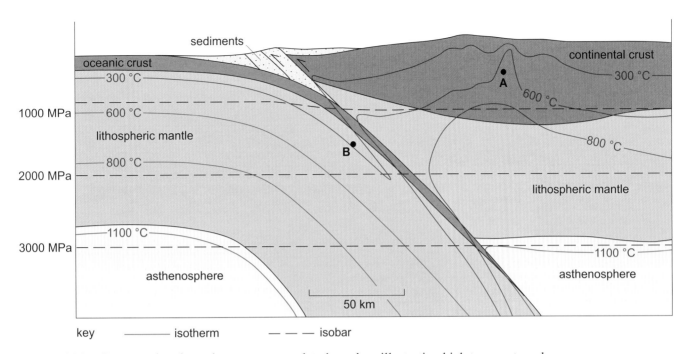

Figure 6.23   Cross-section through a convergent plate boundary illustrating high temperature–low pressure metamorphism (at point A) and high pressure–low temperature metamorphism (at point B). Together these make up a paired metamorphic belt, which runs parallel to the plate boundary.

In general, magmatic provinces are characterised by high heat flow and high geothermal gradients. Moreover, convergent plate boundaries would appear to be unique in that they contain a zone of high geothermal gradients (where the magmas are intruding) adjacent to one where the temperatures have been reduced by subduction (Figure 6.23). The net result is that in addition to the distinctive igneous rocks (Section 5.4) there should also be two roughly parallel suites of metamorphic rocks, one containing relatively high-temperature and low-pressure mineral assemblages, associated with magmatism in the overriding plate, and the other typical of relatively high pressures and low temperatures, associated with subduction of the downgoing slab. This is called a **paired metamorphic belt** and is characteristic of convergent plate boundaries.

Question 6.8

Using Figures 6.21 and 6.23, what sort of geothermal gradient and what metamorphic facies will be present at points A and B on Figure 6.23?

Paired metamorphic belts identified from around the Pacific were first explained simply in terms of the varying geothermal gradients that are associated with subduction zones (Figure 6.23). The presence of blueschist assemblages (at point B) requires not only that very high pressures occurred at low temperatures during active subduction, but also that the rocks were then uplifted rapidly to the surface before they had time to warm up. However, blueschist-facies rocks are not always preserved in ancient subduction zones. In some examples from subduction zones, the initial low-temperature/high-pressure minerals warm and recrystallise as the rocks are brought to the surface, resulting in the loss of blueschist-facies minerals and the preservation of only higher-grade metamorphic facies. Throughout the metamorphic history of any rock in a subduction (or collision) zone, there is a competition between the rate of heating imposed largely by radioactive decay of heat-producing elements and the rate of cooling imposed by the rock being brought towards the surface during exhumation. In the case of blueschists, their high P–low T assemblage is preserved only when the rate of cooling due to exhumation (controlled largely by isostasy) exceeds the rate of heating (controlled by the thermal conductivity of the rocks).

6.5.2   Metamorphism and continental collision

The destruction of an oceanic basin by complete subduction of the oceanic crust that formerly separated two continents is likely to be followed by continental collision and crustal thickening. Such momentous events lead to metamorphism on a regional scale.

■   What is the heat source responsible for regional metamorphism during crustal thickening?

☐   Most of the heat within the Earth is generated by the radioactive decay of U, Th and K, and these elements are most abundant in the rocks of the continental crust, particularly the uppermost crust. So a thicker crust means a steeper geothermal gradient.

The changing geothermal gradient during continental collision can best be understood by considering the effects of thrusting one crustal slice over another, hence thickening the crust. Initially the rocks will heat up, followed by cooling as the rocks are brought to the surface. For a simple model in which a segment of upper continental crust is thrust onto another with a similar geothermal gradient, the immediate result following thrusting is a distinctive 'saw-tooth' variation of temperature with depth (Figure 6.24a; Figure 6.24b, curve 1). At this instant, the pressure in the lower segment has increased, but the temperatures remain as yet unchanged. This clearly cannot last since temperatures of near 0 °C cannot be maintained in what is now well within a thickened segment of continental crust overlain by hot rocks. Moreover, the continuing radioactive decay of U, Th and K, which are now present in larger quantities since there is more upper crustal material in the vertical section, gradually increases the temperature, until, after about 30 Ma, its variation with depth is illustrated by curve 2 of Figure 6.24b. Predictably enough, a rock at a depth of 17 km on Figure 6.24b (near the top of the underlying piece of crust) has had its temperature increased considerably from 200 °C to 800 °C. What is perhaps more surprising is that even in the upper segment of crust the geothermal gradient has also been increased by the emplacement of radioactive crust beneath it.

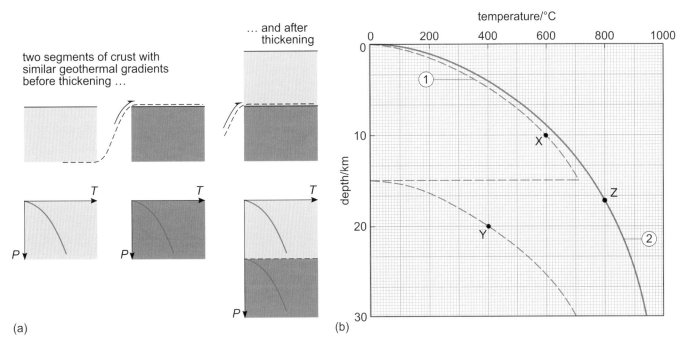

Figure 6.24 (a) Two segments of crust with similar geothermal gradients (shown as curved red lines) before and immediately after crustal thickening. (b) Temperature variation with depth: curve 1 (dashed), immediately after thrusting; curve 2 (solid line) after 30 Ma. Points X, Y and Z are discussed in the text.

Now we shall consider what happens as rocks are brought to the surface. Because of isostasy the thickened crust will be susceptible to exhumation, and rocks must cool as they approach the surface. In some cases, this cooling effect is rapid enough to prevent the geothermal gradient from reaching the equilibrium temperature gradient indicated as curve 2 in Figure 6.24b. If the rocks are exhumed very rapidly after thrusting, then the metamorphic facies appropriate to the *P–T* conditions at points X and Y along curve 1 of Figure 6.24b will be preserved.

■ From Figure 6.21, what are the metamorphic facies of rocks metamorphosed at X and Y?

☐ X lies in the amphibolite facies (600 °C at 10 km) and Y in the greenschist facies (400 °C at 20 km).

Both greenschist and amphibolite facies are represented in the Barrow zones, but what is unusual in Figure 6.24 is that the colder rocks (at Y) occur at greater depths than the hotter rocks (at X). Such a temperature inversion has been observed in metamorphic rocks of the Himalaya and its preservation has been interpreted as evidence of very rapid exhumation, of the order of 5–10 mm per year.

If, as is more usual, the geothermal gradient reaches equilibrium along curve 2 of Figure 6.24b, rocks deep in the thickened crust will reach temperatures of 800 °C or more (at Z), well within the granulite facies. In this case, the diagnostic feature of the collision zones is that the temperatures will increase more rapidly with depth than in crust of normal thickness. Many of the granulite-facies rocks exposed today result from the heating of a crust thickened by collision, and subsequent exhumation of deep crustal levels by isostasy. In the UK, granulites are best exposed in the Lewisian terrane of northwest Scotland, where they formed 2800 Ma ago.

This chapter has mainly been concerned with changes in mineralogy that result from heating rocks at different pressures. However, there are also aspects of metamorphic rocks that relate to deformation, particularly where rocks are folded or faulted. Such mechanical processes form the topic of the next three chapters.

6.6   Summary of Chapter 6

1   Metamorphism is the mineralogical response of a rock to imposed conditions of temperature and pressure that differ markedly from those at which the original rock formed.

2   Broadly, the study of metamorphic rocks comprises: (a) evaluation of the conditions of temperature and pressure to which the rocks were subjected during metamorphism, in the light of experimental data; (b) consideration of the thermal evolution of an area and how that may be explained in terms of its tectonic environment.

3   Metamorphic grade is defined in terms of the entropy change (ΔS) of particular metamorphic reactions: increasing metamorphic grade involves an increase in entropy of the metamorphic system. In general, higher-grade rocks reflect higher temperatures. Prograde reactions are those that take place with increasing metamorphic grade, and retrograde reactions are those that take place with decreasing metamorphic grade. Rates of reaction increase exponentially with temperature, and so equilibrium is more likely to be achieved at higher grades of metamorphism. The presence of a fluid phase will also greatly increase the rates of reaction and its absence will inhibit retrograde reactions.

4   The petrogenetic grid is the pattern of experimentally determined phase boundaries plotted on a phase diagram of temperature against pressure. In principle, once the mineral assemblage in a metamorphic rock has been identified, the petrogenetic grid may be used to estimate the conditions of pressure and temperature under which it crystallised.

5   Dynamic metamorphism occurs within localised areas of intense deformation, such as along fault or shear zones. Contact metamorphism refers to changes in response to heat in the vicinity of igneous rocks. Typically it is not associated with deformation, and the metamorphic grade decreases systematically away from the contact with igneous rocks. Regional metamorphism tends to occur over large areas, and as it is usually associated with collision or subduction zones, the rocks will have been deformed.

6   The particular minerals observed in a metamorphic rock reflect its composition, in addition to the conditions of pressure and temperature and the nature of the fluid phase. Thus different minerals occur in rocks of the same metamorphic grade but of different compositions. The term metamorphic facies is used to embrace all possible mineral assemblages in rocks of different composition that are thought to have crystallised under similar P–T conditions. A metamorphic facies is identified from the observed mineral assemblages, and not from the inferred conditions of pressure and temperature, which are estimated from the petrogenetic grid.

7   The mineral assemblages in metamorphic rocks from different areas can reflect different variations in temperature and pressure and these vary depending on the thermal, and hence tectonic, environment. Geothermal gradients are steep in areas of magmatic activity (e.g. island arcs and mid-ocean ridges), and low at subduction zones. Continental collision and crustal thickening also result in an increased gradient, due primarily to the radioactive decay of U, Th and K within the crust. However, the preserved mineral assemblages depend not only on the geothermal gradient during metamorphism, but also on the rate at which the rocks are brought to the surface.

6.7   Objectives for Chapter 6

Now you have completed this chapter, you should be able to:

6.1 Carry out observations on hand specimens of metamorphic rocks, identify metamorphic minerals and textures and interpret these in terms of the processes involved.

6.2 Interpret simple phase diagrams to infer the conditions of temperature and/ or pressure at which a particular metamorphic mineral, or assemblage, was at equilibrium.

6.3 Map out isograds and use them to assess the relationship of metamorphism to both the intrusion of an igneous magma and a deformation event.

6.4 Evaluate estimates of metamorphic pressure and temperature from metamorphic assemblages to suggest possible tectonic environments in which the metamorphism could have taken place.

Now try the following questions to test your understanding of Chapter 6.

Question 6.9

A fairly common metamorphic reaction is that of:

staurolite + quartz $\rightleftharpoons$ garnet + sillimanite + H_2O

Predict which assemblage has the lower entropy, which is stable at higher grades of metamorphism, and whether this reaction takes place at a staurolite or sillimanite isograd.

Question 6.10

A metamorphosed pelite contains the mineral staurolite in equilibrium with kyanite. Use the following information to estimate the approximate conditions of pressure and temperature during metamorphism:

(a) Staurolite does not form until the temperature exceeds 580 °C in pelitic rocks at pressures greater than 200 MPa.

(b) No granite rocks or migmatites were formed nearby; pelite rocks melt at about 600–650 °C under moderate pressures. You can assume H_2O was present during metamorphism.

(c) The phase relations are as illustrated in Figures 5.3 and 6.4a.

Question 6.11

The inferred metamorphic conditions of temperature and pressure of two rocks from different areas are 200 °C at 800 MPa, and 550 °C at 200 MPa. Using Figure 6.21, for each rock, identify which metamorphic facies it is from, calculate the geothermal gradient (assuming that, for each 1 km in depth, pressure increases by 30 MPa) and suggest in what thermal or tectonic environment it might have been metamorphosed.

Chapter 7   Rock structures

Discussion of continental collision and metamorphism prompts us to investigate deformed rocks. Deformation affects all types of rock – igneous, metamorphic and sedimentary – and is usually caused by differential movement between large bodies of rock, commonly a combination of both horizontal and vertical movements. The features that provide evidence for such movement may be seen worldwide, in exposures of rocks of all ages, and on all scales. Figure 7.1 shows structures at various scales, from a microscopic view to a false-colour satellite image over 150 km across.

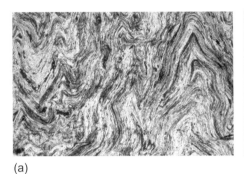

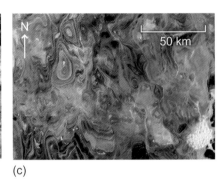

(a)   (b)   (c)

Figure 7.1   Structures on several scales: (a) buckled (folded) layers in a schist seen under the microscope (field of view 5.5 mm across); (b) an exposure (2 m across) of heavily fractured, disrupted limestone from Andalucía, Spain; (c) a false-colour satellite image of folded strata from southern Algeria.

Structural geology is the study of **deformation**, incorporating all the processes that change the shape of rocks to produce rock structures. If our aim is to discover the complete history of any rock, then it is vital to understand how such structures form. Chapter 8 therefore outlines the *theoretical* causes of deformation in rocks, and how factors such as temperature and pressure influence the way in which rocks deform. Chapter 9 examines in more detail how different structures are produced in practice when rocks deform. First, however, this chapter introduces you to the five main structures you are likely to see in the field: joints, faults, folds, cleavage and schistosity. To describe them, you will need to become familiar with some of the terms used in structural geology.

7.1   Brittle deformation and joints

One important way in which rocks can deform involves *fracture*. When a material fractures, it splits into discrete blocks that may or may not move relative to one another. This type of behaviour is called **brittle deformation**. A glass behaves in a brittle way when you drop it onto the floor. Individual pieces of glass (usually rather small) remain coherent and undamaged, but fractures separate the individual pieces, which generally become displaced one from another across the floor. Brittle structures are common in rocks; the main brittle structures are *joints* and *faults*.

Joints are fractures in rocks that show no appreciable lateral displacement between the two sides of the fracture. They are often the most prominent structure visible in rock exposures, irrespective of the rock type (Figure 7.2a and b). They usually take the form of more or less planar surfaces cutting right through the rock, although they can also be curved or irregularly shaped surfaces. Individual joints can frequently be traced for many metres. Joints commonly occur in groups of several parallel or almost-parallel fractures, known as **joint sets**. The spacing between adjacent joints in a set can vary from millimetres to several metres, with some joint patterns visible from space (Figure 7.2c).

Figure 7.2   Examples of joints: (a) well-bedded sandstones and shales, with spaced joints in the sandstones approximately at right angles to bedding. Cliff is about 5 m high; (b) Chequerboard Mesa, Utah, USA: bedding is near-horizontal, while steep joints are picked out as weathered runnels; (c) traces of joint sets and faults, which show up in this false-colour image from the Landsat 7 satellite of part of Arnhem Land in Northern Territory, Australia.

Rocks usually contain more than one joint set. In massive igneous rocks, joints occur both parallel to and at right angles to the surface of an intrusion. In layered rocks, joints may form both parallel and perpendicular to the layers (bedding or foliation), but the perpendicular set will be more noticeable (Figure 7.2a). Commonly, joints intersect on the bedding surface in an X-like pattern, often called a **conjugate** pattern.

■   Look carefully at the joints in the sedimentary layers in Figure 7.2a. How are the joint patterns in the prominent, pale sandstone beds different from those in the darker shale beds?

□ In the sandstones, well-developed joints can be seen at right angles to the bedding. These are almost completely absent from the shale beds, where only the most prominent joints in the sandstones continue into the adjacent shales.

Often where different rock types are interbedded, some lithologies are well jointed, whilst the joints are less well developed in others. Joint development must therefore be dependent in some way on the physical properties of rocks within the succession.

Since joints are gaps and cracks in the rock, they form obvious fluid pathways. Fluids that flow through these cracks may precipitate minerals as veins, so many joints are filled with thin seams of white, crystalline minerals, commonly calcite or quartz (Figure 7.3).

Figure 7.3   This limestone in Switzerland has developed a conjugate set of fractures, mainly joints, which were later infilled with white calcite to form a network of veins.

7.2   Faults

Faults are discontinuities between rock masses that have been displaced a *measurable distance* relative to each other. Whilst joints separate blocks of rock that may have merely moved apart, faults are recognised by horizontal or vertical displacement between two blocks, or more typically a combination of both directions. Major faults, with displacements between several metres and hundreds of kilometres, are represented on geological maps as bold lines, which may be solid (observed) or dashed (inferred). Faults with small displacements (millimetres to metres) are much more common than those with large displacements, and are seldom shown on geological maps.

7.2.1   Hanging walls and footwalls

The two displaced blocks of rocks separated by the fault plane are known as the **hanging wall** and the **footwall** (Figure 7.4b and c). These useful terms are old miners' names – as miners stood on an inclined vein working ore, they had their feet on the 'wall' underneath them whilst another block of rock was overhanging their heads. The footwall is therefore the body of rock *underneath* the dipping fault plane, and the hanging wall is the body of rock *above* the fault plane. As long as the fault plane is not vertical, we can always recognise a hanging wall and a footwall.

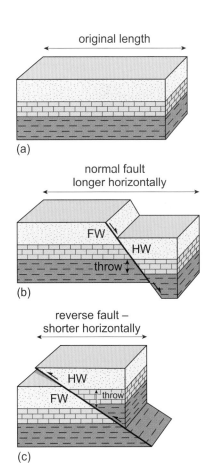

Figure 7.4   Diagram showing how normal faults allow the Earth's crust to lengthen horizontally, whilst reverse faults allow it to shorten horizontally: (a) undeformed strata; (b) normal fault; (c) reverse fault. HW = hanging wall; FW = footwall. The half-arrows indicate the motion of the block on which they are drawn, i.e. the hanging wall in both (b) and (c). The half-arrow in (b) shows slip down the fault plane.

These terms are helpful in describing the movement sense of faults. In normal faults, the hanging wall is the **downthrown** block; it has been displaced (relative to the footwall) *down* the dip of the fault plane. By contrast, in reverse faults the hanging wall has been **upthrown**: it has been displaced *up* the dip of the fault plane (Figure 7.4). As these faults have slipped either up or down the dip of the fault plane, they are known as 'dip-slip' faults.

■   Do (a) normal faults, and (b) reverse faults allow a body of rock to lengthen or to shorten horizontally?

☐   Since the hanging wall of a normal fault has been relatively displaced down the dipping fault plane, it must also have moved *outwards*. This means that the horizontal dimension must increase, so the body of rock must have lengthened horizontally (Figure 7.4b). The opposite is true for reverse faults; upward displacement of the hanging wall must move it *inwards*, meaning that the body of rock has shortened horizontally (Figure 7.4c). The difference between normal and reverse faults is best summarised as follows:

> For normal faults, the hanging wall block has slipped down the fault plane, which therefore dips towards this downthrown block. For reverse faults, the hanging wall block has slipped up the fault plane, which therefore dips towards the upthrown hanging wall block.

In the very rare cases that a fault plane is exactly vertical, a hanging wall or a footwall cannot be defined so, strictly, a vertical fault cannot be classified as normal or reverse (though if there is evidence for horizontal relative displacement, it could be a strike-slip fault). Most strike-slip fault planes are very steep, and because the two rock walls slip horizontally past each other, distinguishing the footwall and hanging wall is not so critical. It is better to describe the relative displacement of one side of the fault viewed from the other, either to the right or to the left, with the terms **dextral** and **sinistral**, respectively (Figure 7.5).

■   If a strike-slip fault was described as dextral when viewed from one side, would it be termed sinistral if you hopped across the fault trace to view it from the other side?

☐   No, the relative displacement would be dextral in both cases.

Fault planes are not often exposed, so the amount or even the direction of dip may be obscure. In such cases, it is generally easiest to consider the fault to be vertical. By convention, vertical faults are assumed to be 'normal' unless there has been demonstrable strike-slip movement.

7.2.2   Investigating fault motion

Certain aspects of faults provide useful information on how the rock strata have been deformed:

•   the dip amount and the strike orientation (called the **trend**) of the fault plane, in degrees or compass directions

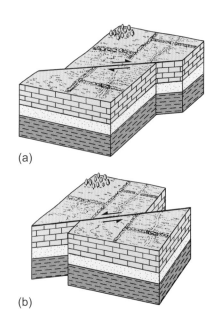

(a)

(b)

Figure 7.5   Block diagrams of strike-slip faults. (a) dextral strike-slip fault; (b) sinistral strike-slip fault.

- the amount of movement between the two sides of the fault (often simply the vertical component of movement, or **throw**), in metres
- the sense of this movement (combinations of the terms dip-slip or strike-slip, normal or reverse).

Direct determination of these properties is rarely possible, as fault planes are seldom exposed. However, the trace of a fault across the land surface approximates to its strike, as will the fault line on a geological map, provided the topography is not too rough. Fault plane dips can be estimated from how they intersect the topography (i.e. steep faults will cut across contours whereas horizontal faults will run parallel to them). Most faults slip intermittently (causing earthquakes) by a few centimetres or metres (at most) at a time. If a **fault scarp** is formed, because a fault cuts the surface, the elevated ground on one side will tend to erode more quickly than the downthrown side. Between slip events, the scarp may well be worn away, exposing older strata juxtaposed against younger rocks on a level surface. **Offsets** of either landscape or geological features may provide clues to the sense of motion on the fault. The throw of a fault, however, can only be determined accurately with good subsurface information (such as that depicted in an accurate cross-section). If the stratigraphy of an area is known in detail, the throw on a fault can be estimated by considering which strata occur on either side of the fault at outcrop. Figure 7.6b shows that, for a steep fault in

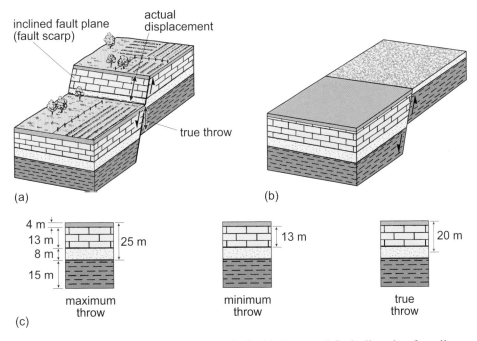

Figure 7.6   Estimating the throw on a fault. (a) A normal fault directly after slip; (b) outcrop pattern of the fault in (a) after erosion has worn down the elevated land surface behind the fault scarp; (c) three sketches of the stratigraphic column for the strata in (a) and (b), comparing two possible estimates of the fault throw with the true throw as shown in part (a).

an area with flat-lying strata, the younger rocks lie on the *downthrown* side of the fault. This is a useful rule of thumb, provided the fault is not cutting structurally complex (e.g. overturned) strata. If the thicknesses of the strata in the area are known, the throw can be estimated by calculating the vertical distance between the two units exposed on either side of the fault, measured on a stratigraphic column.

Question 7.1

Will an estimate based on the stratigraphic separation of the two units on either side of the fault give an exact value for the throw on the fault? Explain your answer.

Generally, an estimate of the throw will be given as maximum and/or minimum possible values: for the fault in Figure 7.6, the maximum throw would be measured from the top of the younger unit to the base of the older unit, while the minimum estimate would use the base of the younger unit and the top of the older unit. The method is more precise if two boundaries between different units are juxtaposed across a fault.

Question 7.2

Study the fault X–X′ trending WNW to ESE just to the north of Hornsleasow Farm (123322) on Figure 7.7.

(a) What is the throw on this fault at (135322), where the base of the Hampen Formation lies almost exactly opposite the top of the Birdlip Limestone Formation? (Measure the thickness of strata from the generalised vertical section, rather than trying to add up the variable thicknesses to the right of the column. Use the vertical scale bar provided in the key for your calculation.)

(b) What is the throw on this fault at (101334)? You can assume that at this point on the fault, the top of the Bridport Sand Formation lies exactly opposite the midpoint of the Salperton Limestone Formation.

(c) Would you ascribe the difference in your answers to (a) and (b) to uncertainties involved in the estimate, or is it more likely to represent a real variation?

All faults have a finite length. At the end of the fault, where it dies out, this must mean that the throw on that fault has become zero. Many faults have a form similar to that shown in Figure 7.8, with a maximum throw near their midpoint, decreasing towards the two fault tips.

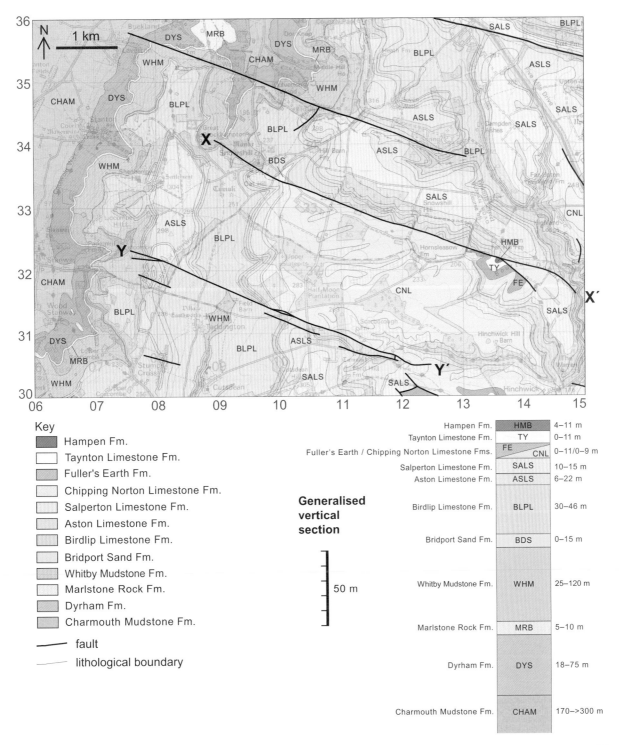

Key
- Hampen Fm.
- Taynton Limestone Fm.
- Fuller's Earth Fm.
- Chipping Norton Limestone Fm.
- Salperton Limestone Fm.
- Aston Limestone Fm.
- Birdlip Limestone Fm.
- Bridport Sand Fm.
- Whitby Mudstone Fm.
- Marlstone Rock Fm.
- Dyrham Fm.
- Charmouth Mudstone Fm.

—— fault
—— lithological boundary

Generalised vertical section

Hampen Fm.	HMB	4–11 m
Taynton Limestone Fm.	TY	0–11 m
Fuller's Earth / Chipping Norton Limestone Fms.	FE / CNL	0–11/0–9 m
Salperton Limestone Fm.	SALS	10–15 m
Aston Limestone Fm.	ASLS	6–22 m
Birdlip Limestone Fm.	BLPL	30–46 m
Bridport Sand Fm.	BDS	0–15 m
Whitby Mudstone Fm.	WHM	25–120 m
Marlstone Rock Fm.	MRB	5–10 m
Dyrham Fm.	DYS	18–75 m
Charmouth Mudstone Fm.	CHAM	170–>300 m

50 m

Figure 7.7   Extract from the geological map of the Moreton-in-Marsh area, including the relevant part of the generalised vertical section.

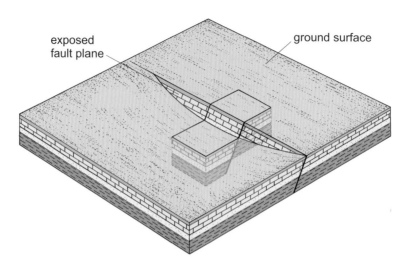

exposed
fault plane

ground surface

Figure 7.8   Block diagram illustrating the way in which a fault can die out at either end. The ground sag in the middle has been exaggerated for clarity, and need not be symmetrical. The inset block corresponds to Figure 7.6a.

As well as estimating the magnitude and sense of faulting, deducing *when* the fault moved can give insights into the geological history of an area. The following general rule can be applied:

The age of a fault must be *younger* than the *youngest* bed (or other geological feature) it cuts, and *older* than the *oldest* bed (or any other geological feature) that is unaffected by the faulting. In other words, the youngest rocks or features that are cut by a fault give the *maximum* age of the faulting, while the oldest rocks or features unaffected by a fault give the *minimum* age of faulting.

7.2.3   Horsts and grabens

The combined effect of one or more faults can produce other structural features. The map extract in Figure 7.7 shows several sub-parallel faults trending generally WNW to ESE to the north and south of Hornsleasow Farm. Here, the downthrow of the adjacent faults X–X′ and Y–Y′ is on opposite sides. The net effect is to produce a *downthrown* block or **graben** between them. A graben is the structure formed when a block of crust is downthrown between two parallel (or near-parallel) faults. Figure 7.9a and b illustrate this for the Hornsleasow area.

A **horst** structure is formed when a block is *upthrown* relative to the surrounding land surface between two parallel (or near-parallel) faults, as shown in Figure 7.9c. A small horst block can be observed between (100313) and (105311) on Figure 7.7.

Many grabens are bounded on one side not by a single fault, but by several more or less parallel faults that all downthrow in the same direction, producing a pattern called step faulting (Figure 7.9d).

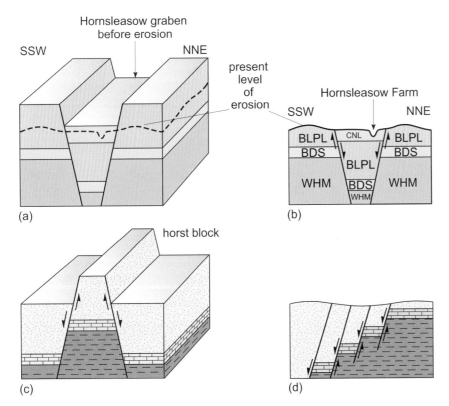

Figure 7.9   (a) Block diagram of Hornsleasow graben before erosion. (b) Cross-section of Hornsleasow graben as it now appears. (See Figure 7.7 for key.) (c) Block diagram of a horst before erosion. (d) Step faulting.

Activity 7.1   Describing and exploring faults

This activity gives you the opportunity to recap on what you have learned about faults in Books 1 and 2, and then to explore how faulting affects outcrop patterns.

7.2.4   Faulting and outcrop patterns

In areas where the Earth's crust is being shortened horizontally, brittle deformation is dominated by low-angle (<45°) reverse faults known as thrust faults (Figure 7.10). In areas of simple stratigraphy, the basic rule for normal faults – that younger rocks lie on the downthrown side – also applies to reverse and thrust faults. However, although it is common for thrusts to emplace older rocks over younger strata, this is not always the case. Thrust faults typically occur in highly deformed regions, with extensive folding and disruption of strata, so a thrust cutting through such complex structures will not necessarily emplace older rocks above younger; much depends on how disrupted the strata were before the fault cut through them.

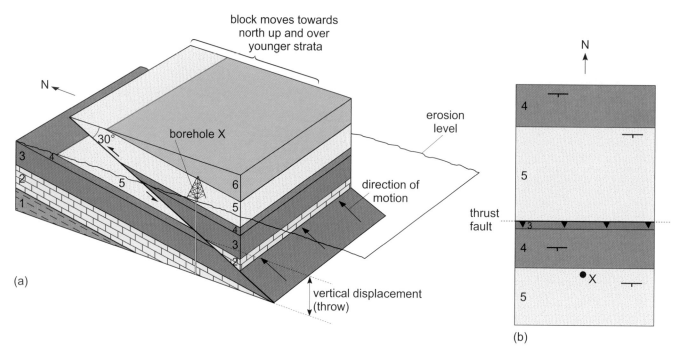

Figure 7.10   (a) Block diagram of a low-angle reverse (thrust) fault in gently dipping strata. The thrust plane dips at 30° southwards: the strata have a slightly shallower dip. In this case, the thrust has carried older strata (e.g. layers 3 and 4) upwards and northwards over younger strata (e.g. layers 4 and 5). (b) Outcrop pattern (in map view) resulting from the thrust after erosion has reduced topography to a level plane. Note that the thrust line is ornamented with triangular teeth on the hanging wall, which is the convention on many tectonic or structural maps. Note also that in (a) younger rocks on the northern side of the fault line (layer 5) lie adjacent to older rocks (layer 3) on the southern side, so the downthrow is to the north. Another important effect of the thrust is that on the plan or map view the stratigraphic succession (layers 4 and 5) is repeated either side of the thrust. This also occurs in the vertical sense, since a borehole at locality X would encounter layers 3 and 4 twice because they are repeated below the fault plane.

Figure 7.11a is a small-scale, simplified geological map of part of northern Montana, showing the eastern edge of the Rocky Mountains. This zone contains numerous thrust faults, so only a few major thrusts are shown on this map, among them the Lewis Thrust.

■   Bearing in mind the mountainous terrain, what feature of the trace of the Lewis Thrust strongly implies that it is a low-angle surface, rather than a reverse fault with a steep dip?

☐   The trace of the fault in map view is highly sinuous, which suggests it is hugging the contours of mountainous topography. This is characteristic of a low-angle surface, such as a thrust or an unconformity.

Southeast of the Lewis Thrust, several thrust faults run parallel to each other (and the mountain front). These faults all dip towards the west. Study the outcrop patterns around these thrust faults, and answer Question 7.3.

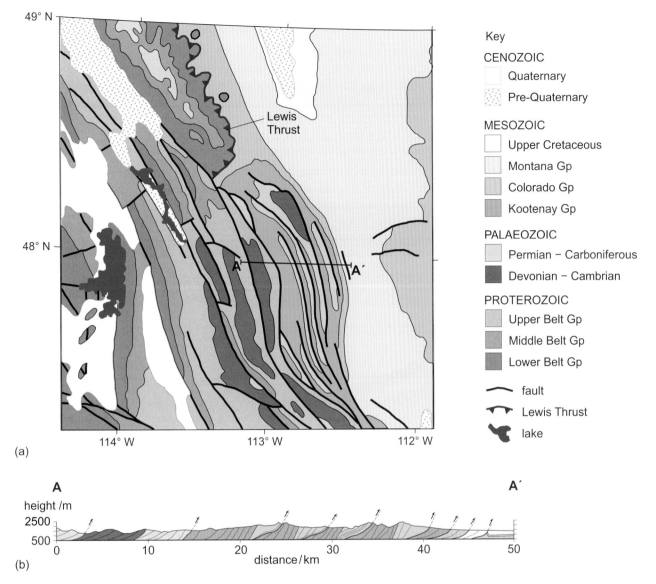

(a)

(b)

Figure 7.11 (a) Simplified geological map of part of northern Montana, USA, showing part of the eastern mountain front of the Rockies. (b) Sketch cross-section along line A–A′ in part (a). Major thrust faults are highlighted in red; thin grey lines are a rough indication of the orientation of strata within the thrust sheets. Colour scheme is identical to that in part (a). No vertical exaggeration (vertical scale = horizontal scale).

Question 7.3

Briefly explain what *two* aspects of the outcrop patterns of strata associated with the parallel faults in Figure 7.11a indicate that they are thrust faults. (You may want to refer to Figure 7.10.)

Each thrust fault shown in Figure 7.11b emplaces older rocks on younger (Proterozoic on Devonian–Cambrian; Proterozoic on Cretaceous; Carboniferous on Cretaceous). The thrusts produce this regular pattern because the strata at the edge of the Rockies have not been contorted as much as the rocks in the interior of the mountain belt.

All faults displace bodies of rock, and in many cases disrupt outcrop patterns. A simple example might be the offsets caused by sideways (i.e. horizontal) motion on strike-slip faults (Figure 7.5). You may find these faults referred to as transcurrent, wrench or tear faults, and any motion on one of these faults will wrench apart surface features, such as fences, streams and roads. However, if the land surface in Figure 7.5 were stripped down to the bedrock, motion on neither the dextral nor sinistral strike-slip fault would disrupt the outcrop pattern, because the strata are horizontal. Although strike-slip faults can have huge displacements (hundreds of kilometres), so long as the slip is parallel to the strike of the beds cut by the fault, the outcrop patterns will not be offset. On the other hand, if a strike-slip fault cross-cuts dipping strata at an angle to their strike direction, then an offset will be generated at the surface with only a small displacement (Figure 7.12a–c). The greatest offset will be produced by a fault cutting the strata perpendicular to their strike, and hence parallel to the dip direction: a **dip-parallel fault**.

Figure 7.12   Pattern of rock outcrop resulting from a sinistral strike-slip fault (F–F') perpendicular to the strike of dipping beds: (a) prior to fault movement; (b) horizontal movement offsetting beds; (c) outcrop pattern (i.e. map view) resulting from fault motion.

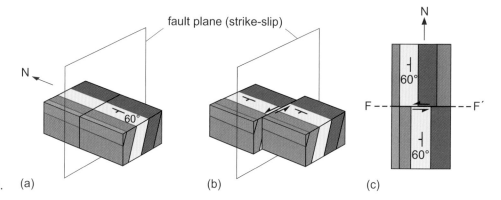

Similar offset patterns can also be generated by both normal and reverse faults, even if the motion on them is purely *dip-slip* (i.e. more vertical displacement), and there has been no strike-slip movement. Figure 7.13a–c illustrates the pattern resulting from the movement on a normal fault that cuts a series of steeply dipping strata. A similar pattern is produced where a reverse fault cuts across the bedding strike (Figure 7.14a–c). It would not be possible to determine whether the offset had been caused by a normal or a reverse fault from simply observing these outcrop patterns presented on a geological map, without additional information (e.g. the dip direction of the fault plane).

Figure 7.13   Pattern of rock outcrop resulting from a normal fault (F–F') perpendicular to the strike of dipping beds: (a) prior to fault movement; (b) vertical movement offsetting beds (wavy line on upthrown block shows surface after erosion); (c) outcrop pattern (i.e. map view) resulting after erosion of both blocks to a common topographic level. Note short 'tick' marks on downthrown side of the fault.

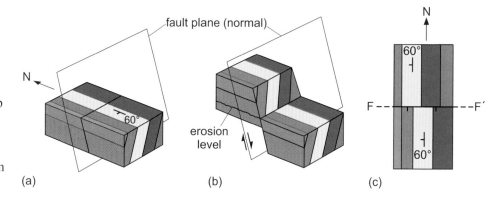

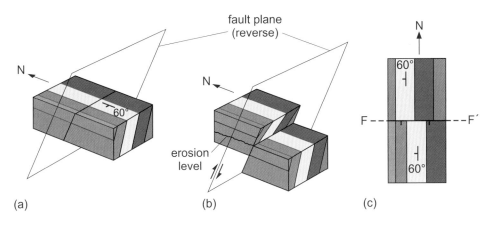

Figure 7.14   Pattern of rock outcrop resulting from a reverse fault (F–F′) perpendicular to the strike of dipping beds: (a) prior to fault movement; (b) vertical movement offsetting beds (wavy line on upthrown block shows surface after erosion); (c) outcrop pattern (i.e. map view) resulting after erosion of both blocks to a common topographic level. Note short 'tick' marks on downthrown side of the fault.

Let us now consider the types of outcrop pattern that would result from a fault cutting the dipping strata in an orientation parallel to the strike (i.e. a **strike-parallel fault**). Here, the important factors that fundamentally affect the outcrop pattern are the angle of dip of the fault plane, and whether the fault plane dips in the same direction as the stratigraphic dip, or opposite to it. Instead of producing an offset outcrop pattern, the outcrop width of the dipping beds cut by the fault may apparently be altered, whilst in some cases beds may be repeated at the surface, or else removed altogether. A good example of the type of repetition and narrowing of outcrop that can result from a strike-parallel reverse fault is given in Figure 7.10b.

Question 7.4

How might the outcrop pattern appear after erosion to the same level if the amount of vertical movement (throw) shown in Figure 7.10a had been half as much again as the amount illustrated? (A simple sketch of the outcrop pattern may help.)

It is most important to realise that commonly, in many real geological situations, faults are neither perpendicular nor parallel to strike, and most cut the strata at an angle that lies somewhere between the dip and strike directions. This makes distinguishing types of fault from outcrop patterns even more difficult, since either vertical or horizontal displacements may produce similar offsets. Only where vertical features (e.g. dykes and vertical beds) are demonstrably displaced by faults can we be certain that some horizontal (i.e. strike-slip) fault displacement has occurred, and measure it on the map.

The important differences in map outcrop pattern that should enable you to distinguish strike-slip and dip-slip faults are as follows:

- Dip-slip faults (i.e. normal or reverse faults) typically have steeply dipping (but not usually vertical) fault planes. When strata dipping in opposite directions (e.g. fold limbs) are cut by dip-slip faults, the predominantly vertical movements produce an outcrop pattern where geological boundaries are displaced in opposite directions.

- Strike-slip fault planes are generally vertical (or nearly so), so their traces cut straight across topography. All geological boundaries on the same side of a strike-slip fault will be displaced in the same direction.

7.2.5   Constructing accurate cross-sections

Geological cross-sections are an invaluable aid to geologists investigating the structure of an area because they add another dimension to the geological map, and allow interpretation of the subsurface geology. Sophisticated geophysical techniques can be used to probe the subsurface structure (e.g. seismic, magnetic, or gravity surveys), but the basic skills of combining surface observations and data with information on local stratigraphy still underpin the construction of any good cross-section. Whenever you construct an accurate cross-section on paper (e.g. Activity 7.2), you should use the approach given in Box 7.1.

Box 7.1 How to construct an accurate cross-section

Your section should show the underground geology to a depth extending down to sea level, and usually below sea level. The depth to which you can usefully extend cross-sections downwards is usually greater in areas of more complex tilted, folded and faulted rocks.

1   You should first decide what scales are appropriate for the horizontal and vertical axes of your section. If possible, keep the vertical and horizontal scales similar, especially if you want to use the dip information on the map. However, on more complicated sections this may not be practical and, in order to show subsurface detail, some vertical exaggeration may be necessary.

2   Draw the topographic profile first. Label the ends of the section with a grid reference or compass bearing. It is conventional to put the westernmost end of the section at the left and, for exactly north–south sections, to put south on the left.

3   Transfer the points of outcrop of boundaries of stratigraphic units, faults and other geological features from the map to the topographic profile. This can be achieved by using a piece of graph paper laid along the chosen line of section and then marking where the geological boundaries or faults occur. If you lay your marked graph paper along the topographic profile, you can then mark precisely where they occur on the topography.

4   For strata with a constant dip, choose an important stratigraphic boundary and draw this across the whole section if possible, joining up the points where it intersects the topography. This 'marker horizon' will help to construct the overall structure of the beds. Put in other units in the same series of strata by determining their outcrop positions from their intersections with the topographic profile and then drawing them as lines parallel to the first one. Try to ensure that the thicknesses of the individual beds you have drawn lie within the range indicated in the stratigraphic column and, if it is drawn without vertical exaggeration, that they remain constant across your section.

5   Are there faults? If so, determine and indicate the downthrown sides. In which directions do the fault planes dip? Make a note of all these features.

6   Are there one or more unconformities present? Check carefully for unconformable relationships by examining outcrop patterns (e.g. truncated boundaries) and the generalised vertical section. You should draw in all the beds above an unconformity *first*.

7   Where dips of strata vary (through folding or faulting), mark each geological boundary with a line above the topographic profile that indicates the dip angle and direction of the strata at or near that point on the map. Then join up the points where each boundary intersects the topography as before, but using the dips recorded at the surface to guide the form of the subsurface boundary. For

instance, surface dip data may indicate a syncline, so you should draw the subsurface boundary as a synclinal curve. Note that you can only transfer true dips onto the section if there is no vertical exaggeration and the dip direction is parallel to the line of section (see the information on apparent dip following point 10, and Figure 7.15).

8 Determine the downthrow side of each fault. Assume faults are normal faults unless there is clear evidence to the contrary, and draw them dipping steeply towards the downthrow side. Where possible, determine the amount of throw on the fault and show this on the section. Label the direction of displacement by using half-arrows.

9 If there is insufficient information for you to be sure of boundaries at depth, on part of the cross-section indicate this by the use of a question mark.

10 Prepare a labelled key by colour shading and/ or lettering the beds shown on the section. Adding annotations, such as the horizontal and vertical scales, codes for different rock units, fold axial surfaces, and labels for major faults or unconformities, can all improve the cross-section. Dashed lines that extrapolate folds above ground can also help clarify the overall structure. Your section is now complete.

If a vertical cliff through strata is oblique to their true dip direction, the dip of the beds viewed in the cliff face always appears *lower* than the true dip. This **apparent dip** is the amount and direction of dip measured on a dipping surface in any direction other than true dip. A simple analogy is to consider one side of the sloping roof of a house (Figure 7.15a). The ridge of the roof lies along the strike direction of the roof surface, whilst the dip direction and amount can be measured along the gable end of the roof, since this is at 90° to the ridge of the roof (i.e. the strike). Any other line drawn across the roof will not be at 90° to the ridge and so its dip will be an apparent dip and will be less than that of the true dip. As shown in Figure 7.15a, there are any number

of apparent dips, varying from very shallow angles (point a on Figure 7.15a) nearly parallel to the strike direction to angles almost as steep (point d) as the (true) dip.

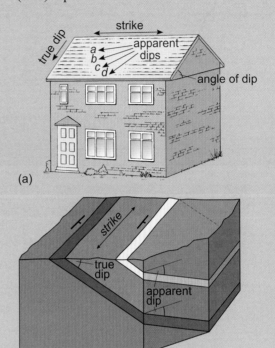

Figure 7.15 House roof analogy to illustrate dip and strike. (a) True dip is parallel to the gable end, in other words at 90° to the roof ridge. Strike direction is parallel to the roof ridge. Apparent dips a, b, c and d. (b) Block diagram with oblique face showing apparent dip of strata. On the right-hand face (parallel to strike), the apparent dip is zero.

A similar situation occurs when a line of section is oblique to the true dip direction of strata marked on a geological map. To depict the strata dips correctly on the section, the true dips must be reduced by a factor that depends on how oblique the line of section is to the direction of true dip. On a cross-section drawn at 90° to the dip direction, the apparent dip of strata is zero. This is why cross-sections should ideally be drawn parallel to the dip (i.e. perpendicular to the strike) of local strata, to best represent the structure.

Activity 7.2   Constructing an accurate cross-section

In this activity, you will put into practice the principles set out in Box 7.1 and construct your own geological cross-section.

7.2.6   Slickensides

As you have seen, it is generally not possible to determine in detail how a fault has moved simply by noting which rocks crop out on either side of the fault. It may be possible to establish, for instance, the downthrown side of the fault; the dip of the fault plane; and whether the fault is normal, reverse, or strike-slip. However, the exact path one wall took relative to the other cannot be established from this information.

Figure 7.16   Steep, aligned grooves (slickensides) on a small fault in the Swiss Alps. The weaker, sub-horizontal ridges may mark where part of the hanging wall block stuck for long intervals between distinct slip events (i.e. earthquakes) on the fault.

Clues to the movement direction lie along the fault plane itself. Many fault planes show a polished surface, called a **slickenside surface**, containing either aligned grooves or elongate crystalline fibres, commonly of quartz or calcite. These grooves or fibres are aligned with the movement direction of the fault (Figure 7.16). They developed when the fault was active: mineral fibres form when the walls are held apart by mineral-rich fluid under pressure, grooves form where they are not. The grooves themselves are called **slickensides**, whilst the fibrous mineral growths are often called **slickenfibres**. Slickensides and slickenfibres give an accurate measure of the direction of slip of the fault. However, the movement history of the fault might be a protracted one and, generally speaking, slickensides and slickenfibres record only the last (or one of the last) episodes of movement, so although they provide an accurate movement direction they rarely provide a complete record of movement. Grooves and fibres associated with real faults commonly show some degree of **oblique slip**, such as the ones in Figure 7.16. In such cases, relative displacement of the two walls is neither exactly up- or down-dip nor along the strike, but somewhere in between.

7.2.7   Detachments

Faults usually cut across beds in sedimentary successions. However, a fault plane can lie parallel to beds in its footwall, or its hanging wall, or both. Where this happens, the direction of downthrow relative to the dip of the fault cannot be easily worked out, so it may be impossible to call them normal or reverse. Such flat-lying faults are often called **detachment faults** (or simply detachments). Detachment surfaces are often only recognised as surfaces of displacement because they contain slickensides or slickenfibres.

7.2.8   Fault tip lines and tip arrays

Faults are three-dimensional structures, and fault planes extend laterally through rocks as well as penetrating upwards and downwards. They are never infinite: the amount of displacement decreases away from a maximum, towards the point at which displacement dies out altogether (Figure 7.8). The locus of all the points on a fault plane where the **fault displacement** dies away to nothing is called the **fault tip line**. The point where the fault tip line intersects another surface, such as the ground surface or a displaced geological horizon, is called the **fault tip** (Figure 7.17a).

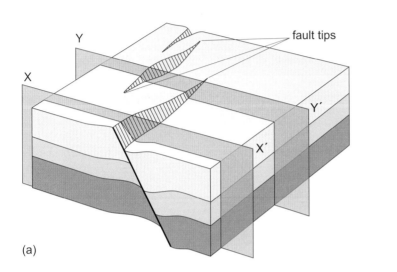

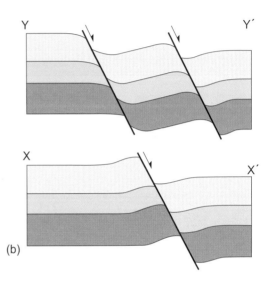

Figure 7.17   Fault tips and fault arrays. (a) A block diagram showing an array of normal faults, which together allow deformation of a large block of rock. Note how the individual faults overlap in plan view (i.e. they run parallel for some of their length). The fault tip line intersects the displaced surface at fault tips. (b) Two cross-sections, X–X′ and Y–Y′, have been drawn through different parts of the block diagram. The number of faults in each cross-section is different, but the amount of horizontal displacement is the same. (Note: The folds shown near to the faults are illustrative of those seen in the hanging walls and footwalls of real major normal faults.)

Figure 7.17 shows schematically how faults of finite size develop together to allow displacement on a large scale. The displacement on each individual fault dies out, but displacement is picked up by a second fault that overlaps the first. In Figure 7.17a, there are three such overlapping normal faults. Even though each individual fault is only of finite length, the overall displacement is maintained across the body of rock, as can be seen from the uniform rectangular shape of the block. Multiple faults with parallel or sub-parallel strike directions are called **fault arrays**.

Activity 7.3   Exploring faults further

This activity builds on the skills you practised in Activity 7.1 relating to faulted outcrops, and includes a short video sequence on DVD introducing some more advanced aspects of faulting as a lead-in to Chapter 8.

7.3 Shear zones and ductile deformation

Outcrops of metamorphic rock, in particular, sometimes show a structure known as a shear zone (Figure 7.18). Shear zones are fault-like structures in which there is evidence of displacement and deformation, but no brittle failure. In shear zones, displacement is distributed within the rock rather than taking place across a single plane, so offsets of marked layers across the zone are not as clear as those across brittle faults. Instead, layers in the rock next to the shear zone appear to be smeared into the zone, typically swinging round parallel with it and becoming stretched out. Even if an offset marker layer cannot be traced through the zone, the sense of shear across the zone can be inferred from the sense of deflection of any pre-existing layering or fabric. They occur on all scales, from the microscopic to major structures several kilometres wide running the length of a mountain belt.

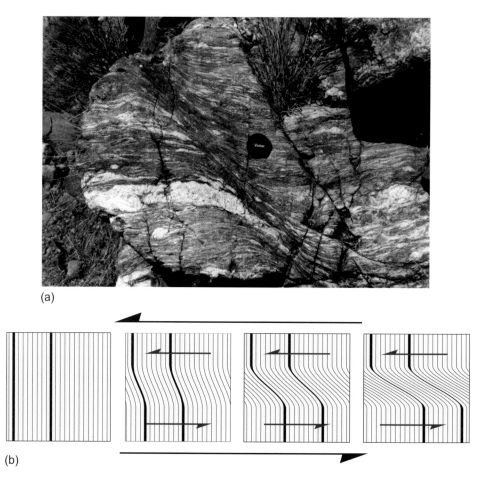

(a)

(b)

Figure 7.18 (a) Photograph of a shear zone within deformed gneiss from southern Spain. Note how the foliation in the rock sweeps clockwise into the zone, until it is almost parallel to the zone itself. (b) Series of sketches to show progressive shearing of a rock with a fabric (black lines). Both the deflection of this fabric, and the offset of black marker layers, indicate that this zone has a sinistral sense of shear. Note that the marker layers become thinner within the shear zone as they are stretched.

Deformation in which material is permanently deformed but does not fracture is called **ductile deformation**. Shear zones are therefore the ductile equivalents of brittle faults. Shear zones are much more common in metamorphic and plutonic igneous rocks, which formed at depth, than in sedimentary successions, which formed at the surface, so in part they can be considered as the deep-seated equivalent of faults.

7.4   Folds

Perhaps the most obvious ductile structures seen in rocks are *folded* rock layers. In a simple anticline, strata form an 'arch' shape, with strata dipping away from each other from the top of the arch. In a syncline, strata dip inwards to form a trough. However, not all folds are so simple, so a number of terms are used to describe their varied geometry (Figure 7.19).

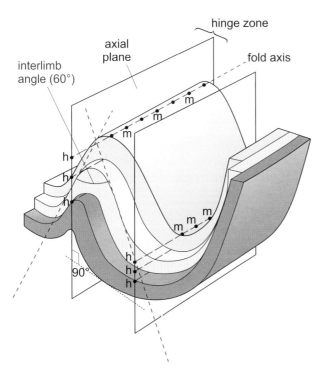

Figure 7.19   Cutaway block diagram illustrating the basic geometry of a simple anticline–syncline fold pair. The hinge line or fold axis (dashed line labelled h), horizontal in this case, joins the points of maximum curvature (m) along each of the bedding surfaces. The axial surface is an imaginary surface that cuts through successive layers along their fold axes; in this series of idealised folds, it is a simple plane called the axial plane. In this example, the axial planes of both folds are vertical, so they are classed as upright folds. Both limbs of the anticline are the same length either side of the hinge line, so the anticline can be described as symmetrical.

7.4.1   Describing folds

Folds are commonly described in **fold pairs** (i.e. an adjacent anticline and syncline together). Each fold has a **limb** on either side, and a **fold axis** (sometimes called a hinge line) that lies along the line of greatest curvature of the folded beds. The **axial surface** (or **axial plane**, if it is a flat, planar surface as in Figure 7.19) is an imaginary surface which passes through the fold axes of successive layers and is a useful conceptual tool for determining whether a fold is *upright* or *inclined*. For instance, in upright folds, such as Figure 7.19, the axial surface is vertical, whereas in inclined folds the axial surface has a dip (e.g. to the north (left) in Figure 7.20). Folds are also often described as being symmetrical or asymmetrical:

- in **symmetrical folds**, the axial surface bisects the angle made by the two limbs exactly (Figure 7.19)

- in **asymmetrical folds**, one fold limb is shorter or steeper than the other (Figure 7.20).

Figure 7.20   Block diagram illustrating axial surface geometry in an inclined asymmetrical anticline and syncline. Here, the axial surface dips steeply northwards so the fold is inclined, but the fold axis (hinge line) is horizontal, as in Figure 7.19. Because the anticline has a long northern limb and a short southern limb, it can be described as asymmetrical. Note that the axial surfaces of both folds are curved, so they cannot be termed axial *planes* in this case.

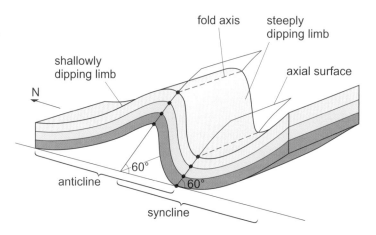

Fold axes are seldom horizontal, as in Figures 7.19 and 7.20; in many cases they are inclined. This inclination of a line from the horizontal is termed the **plunge**, to distinguish it from the *dip* of a plane, and the angle of inclination of a fold axis is termed the **fold plunge** (Figure 7.21).

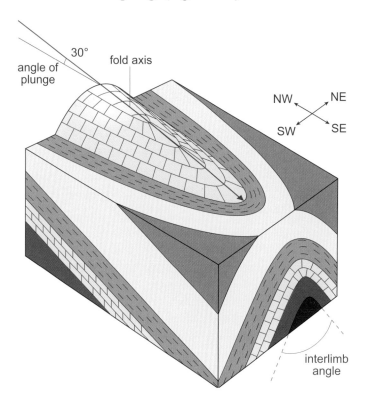

Figure 7.21   Diagram of an open fold plunging gently from the horizontal at 30° towards the southeast.

Folds are typically described in terms of the following:

- The *dip* amount and the *strike* orientation (the trend) of the *fold axial surfaces*, in degrees and a compass direction. For the basal parts of the folds in Figure 7.20, these dip at 60° northwards, and trend E–W.

- The amount and direction of *plunge* of the fold axes, also in degrees and a compass direction. The fold shown in Figure 7.21 plunges towards the southeast at 30° (measured from the horizontal).

- If the fold pair is asymmetric, the *sense of asymmetry* of the fold is expressed as a direction of steepening or apparent overturning of the limbs of the fold. This is very often opposite to (i.e. 180° from) the dip direction of the axial surface. For example, the asymmetric fold pair shown in Figure 7.20 is asymmetric towards the south.

- The angle subtended internally between the two limbs, known as the **interlimb angle**, which may either be expressed in degrees or more usually by a simple descriptive term, defines five main classes:

gentle	180°–120°
open	120°–70°
close	70°–30°
tight	30°–0°
isoclinal	0°

- One of the folded layers in Figure 7.19 has an interlimb angle of about 60° (allowing for perspective), and would therefore be called a close fold. The plunging anticline shown in Figure 7.21 has an interlimb angle of about 90° and is an open fold.

- The general shape of the fold closure, either *rounded* (like a 'U') or *angular* (like a 'V') (Figure 7.22).

(a)   (b)

Figure 7.22   (a) Rounded folds from northeast Scotland; (b) angular folds from Kangaroo Island, South Australia.

■   Are the folds in Figure 7.22b symmetric or asymmetric?

☐   Adjacent limbs in these folds are of unequal lengths, so these folds are asymmetric.

■ How would you classify the folds in Figure 7.23, based on (i) their axial planes; (ii) the plunge of their fold axes (*Hint*: look at the sandstone bed highlighted by the dashed yellow line on the left of the photograph); (iii) their symmetry; (iv) their interlimb angles; (v) their angularity?

Figure 7.23   Folds in bedded sediments, on the west coast of Wales. The view is towards the NNE, from a higher elevation than the top of the cliffs opposite. Also for use in Question 7.5.

☐ The three-dimensional perspective of this photograph gives a better impression of the different elements of these folds than the more usual two-dimensional cliff sections.

(i)   The fold axial planes dip fairly steeply (~70°) towards the right, i.e. ESE, so the folds are steeply inclined.

(ii)   The plunge is trickier to estimate, but on the wave-cut platform at the extreme left end of the near cliff, one sandstone bed forms a tight 'V' shape pointing northeast (away from the observer). The waves have cut through a tight anticline here, and the direction of the 'V' is therefore in the direction of plunge (roughly northeast). The angle of plunge appears to be fairly gentle (10°–30°).

(iii)   Although an entire fold pair cannot be seen, the folds appear to be mildly asymmetric (the western (left-hand) limb of the nearer syncline seems longer than its eastern (right-hand) limb).

(iv)   The near fold has an interlimb angle of around 80°–90° (i.e. it is an open fold), but that of the further fold is less than 70°, so it is a close fold.

(v)   Both folds have mainly straight limbs and moderately angular hinges, so they are moderately angular rather than rounded.

Were we to visit this outcrop, we could measure dips, strikes and plunges for a more precise description of the folds.

Deformed rocks show folds on all scales, so it is useful to be able to describe the dimensions of folds. The width of a fold pair is called the **fold wavelength**; the height of the fold is called the **fold amplitude** (Figure 7.24). These terms are analogous to other waveforms; the wavelength is the lateral distance between adjacent anticlinal (or adjacent synclinal) fold hinges and the amplitude is half the height between adjacent anticline and syncline hinges.

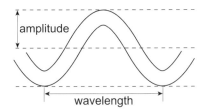

Figure 7.24   Wavelength and amplitude of simple folds.

Question 7.5

Look at the folds in Figure 7.23 and roughly estimate (a) the wavelength, and (b) the amplitude of these folds. (The near cliff is about 20 m high at the synclinal axis.)

Folds are further classified using two terms, with reference to the *dip* of their axial planes (approximating the axial surface to a plane) and to their *plunge* (Figure 7.25). The dip amount and dip direction of the fold axial plane can be measured in the same way as dipping of beds or faults. Folds are described as upright, steeply inclined, moderately inclined, gently inclined, or recumbent, depending on the amount of dip of their axial plane (the horizontal axis of the graph in Figure 7.25). Folds are said to be sub-horizontal, gently plunging, moderately plunging, steeply plunging or sub-vertical, depending on the amount of plunge of the fold axis (the vertical axis of the graph in Figure 7.25). Folds are always described using two terms, for example *upright* folds with *gently plunging* axes.

Folds with almost horizontal (dip < 10°) axial planes are called **recumbent folds**. If the amount of *plunge* of the fold is almost the same as the amount of *dip* of the axial plane, the folds are said to be **reclined**. Reclined folds, therefore, plunge directly down the axial plane, parallel or almost parallel to the dip direction. Recumbent folds may or may not be reclined.

Figure 7.25   A fold classification system. The amount of dip of the axial plane is recorded in degrees on the horizontal axis of the graph and the plunge in degrees on the vertical axis.

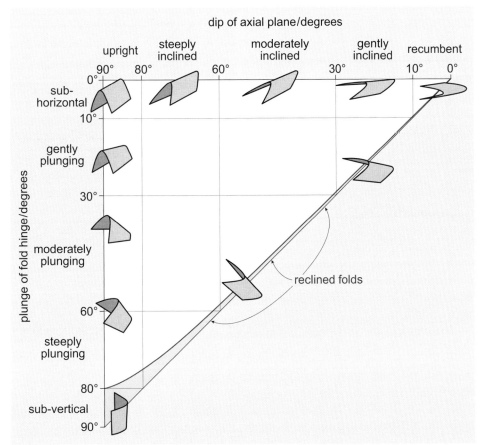

7.4.2   Folds and outcrop patterns

On a geological map of a flat area, eroded folds with horizontal fold axes (Figure 7.26a) show up as parallel belts of outcrop each repeating the strips

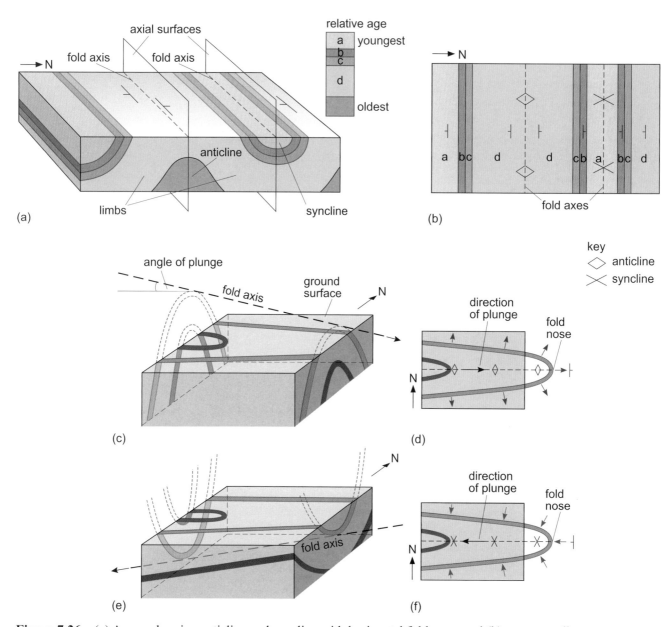

Figure 7.26 (a) A non-plunging anticline and syncline with horizontal fold axes, and (b) corresponding outcrop pattern seen on a map. (c) A plunging anticline with the direction of plunge to the east, and (d) its corresponding outcrop pattern seen on a map. (e) A westward-plunging syncline, and (f) its corresponding outcrop pattern. Note that along the axis of a plunging fold, the bedding dip and the plunge of the fold axis are the same. When drawing sketch maps, it is conventional for geologists to illustrate the direction of plunge using a large annotated arrow as shown in this figure. Another useful symbol is to use paired arrowheads (forming either an X or diamond shape) along the fold axis to show whether the beds dip towards the axis (syncline, (f)) or away from the axis (anticline, (d)).

representing the stratigraphic succession of the folded beds (Figure 7.26b). However, fold axes are seldom horizontal, so outcrop patterns produced by plunging folds are much more common.

Figure 7.26c–d shows an upright anticline with a fold axis gently plunging eastwards, and the resulting outcrop that produces a pattern that *converges* in the direction of plunge. Figure 7.26e–f shows a westward-plunging syncline. Here, the outcrops *diverge* in the direction of plunge.

Wherever there is a plunging fold, the outcrops of the two fold limbs will converge and close, eventually meeting on the fold axis at the fold *nose* (also known as the fold closure). Adjacent anticlines and synclines often plunge in the same direction as each other so, as a result, areas of folded rocks are often characterised by zigzag outcrop patterns, such as those illustrated in Figure 7.27.

Question 7.6

Given that the outcrop patterns of the eastward-plunging anticline (Figure 7.26d) and the westward-plunging syncline (Figure 7.26f) are superficially similar, how would you go about differentiating the two structures on a geological map?

Question 7.7

(a) What type of fold is present at point X on Figure 7.27c?

(b) In which direction does it plunge?

Activity 7.4   Exploring folds and dipping strata

This activity draws together what you have learned about folds in this book and Book 1 by considering how folding affects outcrop patterns on geological maps.

So far, you have considered strata which have been folded into more or less regular shapes. However, anticlines and synclines can plunge at *both* ends, in which case they are known as **domes** and **basins**, respectively (Figure 7.28a and c). These structures can produce circular or elliptical outcrop patterns in map view (Figure 7.28b and d, Figure 7.29 and see also Figure 7.1d).

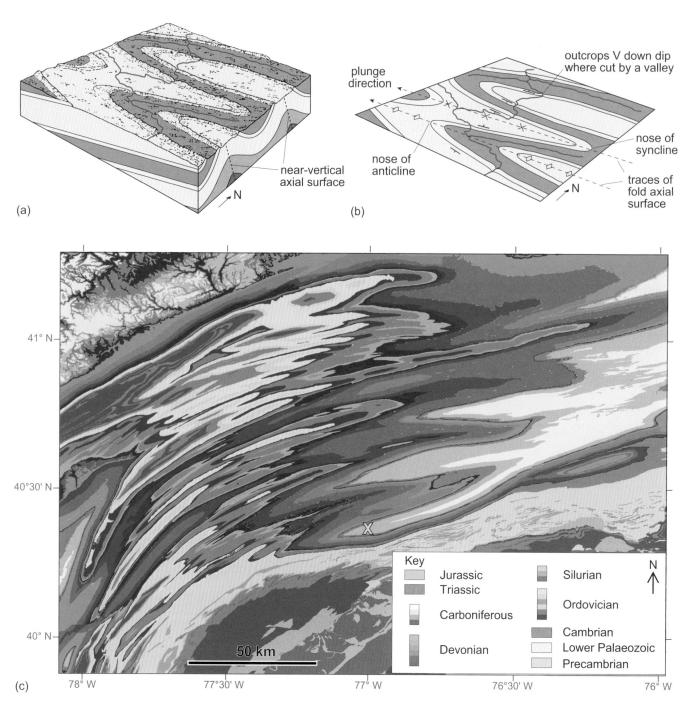

Figure 7.27 (a) Block diagram and (b) corresponding geological map showing the outcrop pattern in an area of folded strata that plunge gently to the west. These folds should be described as upright symmetrical plunging folds (i.e. they have vertical axial planes, limbs of similar steepness and inclined fold axes). Harder beds in the folds form a series of zigzag ridges with steeper scarp slopes facing *inwards* towards the axis of the anticline, and *outwards* from the axis of the syncline. Note also the direction of the V-patterns as the rivers cut across the anticlinal and synclinal axes: V-patterns point inwards towards the axis of the syncline, and outwards from the axis of the anticlines. (c) Outcrop pattern of plunging folds in Pennsylvania, USA. The curving chevron pattern of the folded strata marks the core of the Appalachian mountain belt. Contrast the outcrop pattern in the northwest corner of the map, where sub-horizontal beds are incised by rivers. Point X is for use in Question 7.7.

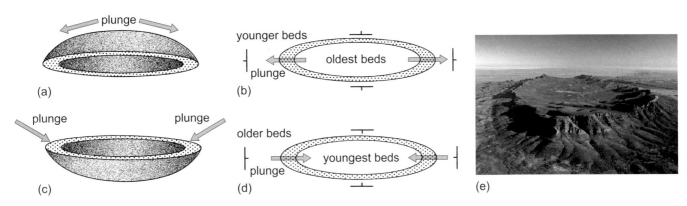

Figure 7.28   (a) An anticline can have the form of an elongate dome when the fold axis plunges in opposite directions at both ends. This produces a dome structure which, when eroded, results in an elliptical outcrop pattern as shown in (b). (c) A syncline with its axis plunging at either end results in a basin structure, which also produces an elliptical outcrop pattern as shown in (d), but with the beds dipping inwards towards its centre. (e) Aerial view of Wilpena Pound, a synclinal basin in South Australia rimmed by resistant strata.

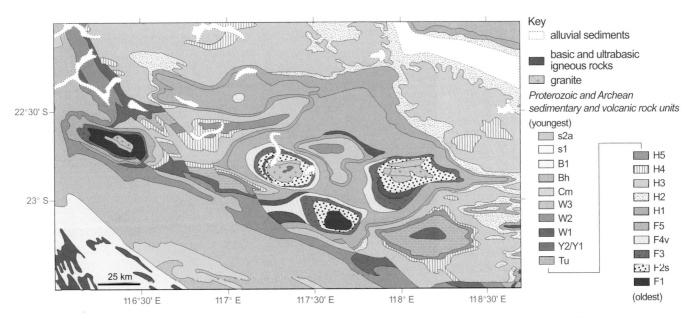

Figure 7.29   Map patterns typical of an area of folded strata, showing domes and basins as distorted elliptical features, from Hamersley Range, northwest Australia.

In some folds, strata do not even dip in opposite directions either side of the fold axis. For instance, in an **overturned** fold, one of the fold limbs has been turned upside down, so that beds on either side of the fold axis dip in the same direction

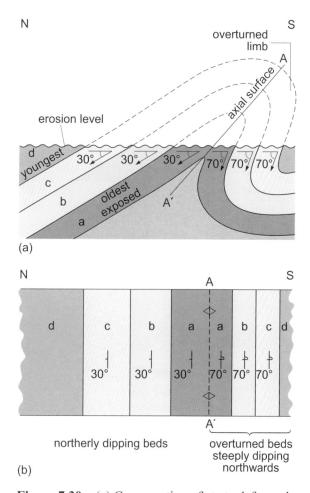

(a)

(b)

Figure 7.30   (a) Cross-section of strata deformed into an overturned fold. The axial surface is inclined and dips northwards, and the steeply dipping beds of the southern limb have been overturned. (Note that the eroded part of the structure is shown using dashed lines above the current erosion level.) (b) Outcrop pattern resulting from erosion of the overturned fold. The beds of both limbs dip in the same direction (northwards), but the dips of the overturned southern limb are shown using the overturned dip symbol. Note that in the case of overturned limbs, the beds young in the opposite direction to dip. (Axial surface of fold is shown as dashed line A–A′.)

(Figure 7.30a). Beds in the overturned limb in this case dip more steeply than those on the other limb, and their dip arrow symbols are shown hooked round the strike line in Figure 7.30b, indicating that the strata are overturned.

Folds must be ductile structures, because beds that are folded are still continuous, with no obvious displacement across any fractures. Many folds indicate ductile deformation of a large volume of rock, caused by tectonic forces acting across a wide region. A word of warning, though – some folds that you may see in rocks, for example those caused by the contortions of flowing lava, or slumped sediments, can be quite common locally. However, such folds reflect local processes rather than large-scale crustal movements.

7.5   Cleavage and schistosity

There is one further very important ductile structure that shows that rocks have been deformed: *rock cleavage*. Cleavage is a foliation found in low-grade metamorphic rocks such as slates. The term *schistosity* is used for foliations in higher-grade metamorphic rocks. In slates, it is only possible to see the individual grains with a microscope, whereas in schists it is sometimes possible to see individual grains with the naked eye.

Cleavage is an alignment of platy or tabular minerals, typically clays and micas. Very often the alignment of fine-grained platy minerals is almost perfect (Figure 7.31) and the rock is then said to possess a **slaty cleavage**. The term cleavage derives from the fact that slates can be split or 'cleaved' along the grain of the mineral alignment. A general term for both the orientation of the mineral alignment and the splitting direction is the **cleavage plane**.

Figure 7.31   A fine-grained rock with a well-developed slaty cleavage. The diagonal tonal banding is the original sedimentary layering, and the fine, vertical lines are the slaty cleavage.

If you study Figure 7.31 carefully, you can see that the slaty cleavage cuts across the original sedimentary bedding. The alignment of platy minerals cannot, therefore, be a sedimentary feature – it must be a response to some process that has affected the rocks after their deposition. In other words, cleavage is a response to deformation. The absence of significant fractures associated with cleavage suggests that cleavage is a ductile structure.

If bedding has been folded, cleavage planes commonly lie parallel, or almost parallel, to the axial planes of folds (Figure 7.32). This is **axial planar cleavage**. If deformation is intense and protracted enough, the same processes that fold bedding planes can also fold cleavage planes. In schists or gneisses, it is common to see a folded foliation (Figure 7.33), frequently accompanied by the formation of new cleavages that are axial planar to the folds of the earlier schistosity. In such rocks, bedding can be very difficult to detect because the growth of new metamorphic minerals masks the original colour or grain size variation. Some metamorphic rocks possess a strong colour banding that may be entirely unrelated to bedding (Figure 7.33).

Figure 7.32 An anticline in bedded rocks in Wales (bedding is picked out by the tonal variation) with an associated, steeply dipping, axial planar cleavage (running from upper left to lower right, parallel to the yellow pen near the centre of the photograph). These structures were formed during the Caledonian Orogeny.

Figure 7.33 Folded foliation in a gneiss, Kinnu, northwest India. The regular light to dark colour banding here is not bedding, but a foliation defined by segregation of metamorphic minerals. View is 11 cm across.

We have described the main types of brittle and ductile structures you are likely to see if you examine outcrops in the field. In the next chapter, you will investigate the processes that cause those structures to form.

7.6   Summary of Chapter 7

1   Rock structures provide an important record of events that have affected rocks of all types, and at all scales.

2   Joints are brittle fractures in rocks that show no appreciable lateral displacement across the fracture. They are commonly seen as planes cutting right through the rock mass, and often occur in sets. They may be marked by veins of minerals precipitated from fluids.

3   Faults are brittle fractures that show a significant lateral displacement between hanging wall and footwall. They can show dip-slip or strike-slip movement, or a combination of both. The three most important categories of fault are normal faults, reverse faults and strike-slip faults.

4   Normal faults act to extend crust horizontally, in some cases creating horst and graben structures. Reverse or thrust faults shorten the crust horizontally.

5   Most faults disrupt outcrop patterns on maps, in many cases producing offsets or features that may give insights into the fault motion. Repetition of strata on a map, for instance, is commonly a result of thrust faulting.

6   Faults typically occur in arrays, which allow deformation to be regional in extent even though individual faults die out.

7   Slickensides and slickenfibres are linear features found on fault planes that enable the orientation of movement to be established.

8   Shear zones are the ductile equivalents of brittle faults, and are often found in rocks that have been deformed at depth.

9   Folds are ductile structures that represent permanent deformations of the original sedimentary bedding or other types of layering. Folds are categorised in terms of their wavelength, amplitude, dip of their axial surface, plunge and shape.

10   Map outcrop patterns provide useful information on the form of folds, especially their plunge and limb orientation. Doubly plunging folds form characteristic 'dome and basin' outcrop patterns.

11   Cleavage and schistosity are alignments of platy or tabular minerals. Both are ductile responses to deformation.

12   Cleavage and schistosity nearly always cross-cut bedding, and nearly always have a close relationship with the geometry of contemporaneous folds.

7.7   Objectives for Chapter 7

Now you have completed this chapter, you should be able to:

7.1 Recognise the main features that show rocks have been deformed.

7.2 Distinguish between joints and faults, folds and cleavage in photographs and field examples.

7.3 Use fault and fold terminology to describe the different types of faults and folds.

7.4 Use outcrop patterns on maps, along with stratigraphic information, to deduce the geometry and nature of different faults and folds.

7.5 Explain the difference between brittle and ductile structures.

7.6 List the main types of brittle structure and ductile structure that you might expect to see in deformed rocks.

Now try the following questions to test your understanding of Chapter 7.

Question 7.8

What is the difference between thrusts and reverse faults?

Question 7.9

Describe the important characteristics of normal faults, thrust faults and strike-slip faults.

Question 7.10

State, giving your reasons, whether joints are brittle or ductile structures.

Question 7.11

Explain whether the shear zone in Figure 7.18a shows sinistral or dextral displacement.

Chapter 8   Rock deformation: cause and effect

We now turn our attention to *process*. Why do rocks deform in certain ways? What controls whether rocks experience brittle or ductile deformation? How do the structures observed in field exposures relate to rock-deforming processes? We start by considering two fundamental concepts – stress and strain.

8.1   Stress in rocks

Stress and *strain* are words in common use: the stresses and strains of everyday life, for instance. However, these terms have precise definitions in physics and rock mechanics. **Stress** is defined as force per unit area and is measured in pascals (Pa) – units that are also used for measuring pressure. Stress becomes greater if the force gets bigger, or if the area over which a given force acts gets smaller. Walking on snow requires broad snowshoes to spread the force (determined by your mass) over a wide area. With stiletto heels, you would sink in at each step: the same force would exert a much greater stress on the snow, because that force would be acting over a much smaller area. Similarly, it's much easier to push a drawing pin into a wall if you press on the flattened side of the pin!

Pressure is the term used when stresses act equally in all directions. An everyday example is atmospheric pressure, where stresses act equally in all directions on us due to the gravitational attraction between the Earth and its atmosphere. Rocks at any point within the Earth are subjected to a similar confining pressure, which is also gravitational in origin and a function of the mass of rock that surrounds that point. Generally speaking, the deeper you go, the greater the confining pressure becomes. Pressure is one of the major controlling factors in metamorphism, as you saw in Chapter 6.

Theoretically, the stresses acting on a volume of rock can be resolved into three components, the three principal stresses, which always act at right angles to one another. Where confining pressure due to the rock overburden is the only stress acting at a point, these three principal stresses are equal (because pressure is defined as stresses acting equally in all directions). In Figure 8.1a, this means that $\sigma_1 = \sigma_2 = \sigma_3$ (the Greek letter sigma (σ) is used to denote stress). Note that, in this diagram, each principal stress is drawn as a *pair* of opposing arrows, because each applied force has an equal force opposing it. An unopposed force would be drawn as a small arrow.

■   Considering σ_1 alone, what would happen to the rock cube if a single, unopposed force acted on it?

□   The unopposed force would push (and in fact accelerate) the cube in the direction of the force, according to Newton's Third Law of Motion: *'For a body at rest or in uniform motion, to every action there is an equal and opposite reaction.'*

Because rocks in the Earth do not accelerate much (if at all), we define a geological stress as an *equal* and *opposite* pair of forces, as depicted in Figure 8.1a. In this figure, all three principal stresses are compressive – they are acting to squash the rock and reduce its volume.

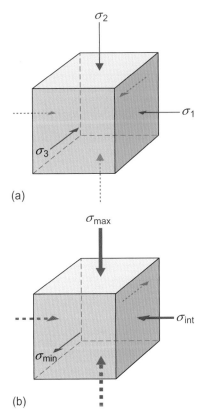

Figure 8.1   (a) The three principal stresses σ_1, σ_2 and σ_3 are always mutually perpendicular, and each stress is represented as a pair of equal and opposite forces. This situation, where $\sigma_1 = \sigma_2 = \sigma_3$, corresponds to the confining pressure at any point in the Earth due to rock overburden alone. (b) In the more general situation, the three principal stresses are unequal (represented by different sized arrows). σ_{max} is the most compressive stress, σ_{min} is the least compressive stress, and σ_{int} is intermediate between σ_{max} and σ_{min}.

Whilst the orientation of each principal stress is established – each is mutually perpendicular to the other two – the magnitudes of the stresses are not. Each principal stress might be either *compressive* or *tensile*; that is, each stress could act to shorten or lengthen the rock *in that orientation*. More generally, and certainly whenever rock is deforming tectonically, the magnitude of the three principal stress components will *not* be equal. One of the three has to be the most compressive and another has to be the least compressive (i.e. less compressive than the other two). The third will have an intermediate value between these two extremes. By convention, unequal stresses are labelled such that σ_{max}, σ_{int} and σ_{min} are the greatest, intermediate and least *compressive stress*, respectively[2] (Figure 8.1b). In real situations, σ_{min} is usually tensile whilst σ_{max} is compressional. However, it is possible for any or all of the stresses to be tensile or compressive, but both σ_{min} and σ_{int} are, by definition, always *less compressive* (and thus more tensile) than σ_{max}. By convention, geologists define compressive stresses as positive, and tensile stresses as negative.

8.2   Strain in rocks

When stresses that are not equal in all directions act upon a body, they tend to displace it, or to deform it, or both. Such stresses generally produce a change in shape. We give the name 'strain' to these changes, so stress produces strain and, therefore, a change in shape. Change in the shape of an object can be measured in different ways, either in terms of changes in length, angle, area, or volume. In turn, the amount of strain is defined as the change in length (or angle, area or volume), relative to the original length (or angle, area or volume). This means that strain is a dimensionless quantity. Strain is simply a change; large amounts of strain mean big changes.

In some cases, strain is linked to stress in a simple way. For example, a spring balance stretches when weights are loaded onto it. The amount that the metallic spring stretches (strain) is proportional to the load (stress) – a fact that is used to calibrate the spring balance.

Within a spring balance, strain is generally *recoverable*. This means that the material returns to its original shape after removal of the applied stress. The same applies when, say, a rubber band is stretched, as long as the force used is small and is not applied too quickly. Recoverable strain is also known as **elastic deformation**, which is why a rubber band is sometimes called an *elastic* band.

■ Is elastic deformation part of brittle or ductile behaviour?

☐ Since the rubber band does not develop any fractures, its behaviour is ductile, not brittle.

---

[2]You may also see the principal stresses labelled σ_1, σ_2 and σ_3. These are directly equivalent to σ_{max}, σ_{int} and σ_{min} by geological convention.

■ When does a rubber band show non-elastic deformation?

☐ If you pull the rubber band hard and quickly, it will probably snap. This is non-elastic deformation, because you cannot get it back to its 'unsnapped' state.

■ When the rubber band snaps, is that brittle or ductile deformation?

☐ Brittle deformation.

Elastic deformation is part of ductile deformation; brittle deformation can never be elastic. Rocks do exhibit elastic behaviour, for example when the seismic waves of a distant earthquake pass through them. Closer to an earthquake epicentre, however, deformation is more likely to be permanent, to both rocks and buildings. You even cause a very small elastic strain every time you stand on a rock exposure!

■ Why don't the structures seen in rocks represent elastic strain?

☐ Rock structures cannot represent elastic strain, because elastic strain is recoverable and the deformation (e.g. folds or faults) is permanent.

After a certain amount of strain, most materials stop behaving elastically and acquire a permanent deformation, which is what geologists observe and describe. Since stresses cannot be measured directly in rocks that were deforming millions of years ago, it is very rare that the exact stress pattern that caused the rocks to deform can be established. Knowing that shape changes must have been produced by stress does not usually allow geologists to *quantify* the parameters that link the two.

There are three fundamentally different ways in which rocks may show strain (Figure 8.2). First, they may move from one place to another. Motion like this, where every point within the rock body moves in an identical way to its neighbour, is known as **translation**. Secondly, a rock body may change orientation, and undergo **rotation** where all points rotate together around some fixed axis. The third type of change is called **distortion**, which is a change of internal shape of the rock. When a rock is distorted, individual points within the rock move relative to one another.

The general term **deformation** is used for all these changes taken together, because most strains involve an interplay of translation, rotation and distortion on a range of scales. In the rock record, it is easy to identify distortion, harder to identify rotation, and usually very difficult indeed to identify translation. Be aware that rocks may be deformed even when they have not been distorted, for example most dipping beds record a deformation.

translation

rotation

distortion

Figure 8.2   The three components of strain: translation, rotation and distortion.

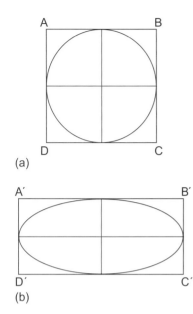

(a)

(b)

Figure 8.3   Distortion of (a) a square block containing a circle into (b) a rectangular block containing a strain ellipse. Note that the areas of the undeformed square ABCD and the deformed rectangle A′ B′ C′ D′ are the same.

8.3   Strain and the strain ellipse

Distortions are commonly described in terms of the amount of shape change of an originally symmetric shape, such as a square or a circle. In Figure 8.3, an undeformed square ABCD containing a circle has been 'squashed' vertically into a rectangle A′ B′ C′ D′. The circle has become an ellipse which, since it records the amount and orientation of strain, is called the **strain ellipse** for that deformation. The ratio of the long axis to the short axis of the strain ellipse, known as its **axial ratio**, gives an indication of the *amount* of strain. The angle the long axis of the ellipse makes with any given reference direction gives the *orientation* of the strain.

■ Look at the square in Figure 8.3a. How would you describe the changes in length of the sides of the square ABCD as it deformed into the rectangle A′ B′ C′ D′ in Figure 8.3b?

☐ The top and bottom are longer than they originally were, and the sides have become shorter.

■ Have any angles changed as the square deformed to a rectangle, either between two adjacent sides (e.g. angle DAB compared with angle D′A′B′) or between the diagonals (e.g. the angle between AC and BD compared with the angle between A′C′ and B′D′)?

☐ Adjacent sides were originally at right angles and are still at right angles. However, the angles between diagonals, originally right angles, have changed.

Generally during deformation there are simultaneous changes in both lengths and angles. Area often changes too, but in Figure 8.3b the rectangle has deliberately been drawn so that it has the same area as the square in Figure 8.3a, so this deformation has not involved a change of area.

8.3.1   Changes in length during deformation

Changes in length of lines or linear features in rocks during deformation are recorded by a measure called the **extension**. The extension, *e*, of any original line is given by the equation:

$$e = \frac{l - l_0}{l_0} \tag{8.1}$$

where l_0 is the length of any line before deformation, and *l* is the length of the same line after deformation.

Numerically, extension is the ratio of the change of length to the original length. Extension is positive if the length of the line increases, and negative if the length of the line decreases. We shall refer to positive extensions as *lengthening* and negative extensions as *shortening*. Extension is commonly expressed as a percentage change in length:

$$\% \text{ extension} = e \times 100\% \tag{8.2}$$

Question 8.1

What is the % extension of (a) the long axis and (b) the short axis of the ellipse in Figure 8.3b? (*Hint*: Use Equations 8.1 and 8.2, and compare the post-deformation lengths of the coloured axes in Figure 8.3b with their pre-deformation equivalents on Figure 8.3a.)

There is a practical difficulty in calculating extensions in deformed rocks. In many cases, the length of some particular feature in the deformed rock can be measured, but how long that feature was *before* deformation is generally unknown. However, in many instances this can be estimated. For example, Figure 8.4 shows three stretched fossils from deformed slates in Switzerland. A geologist could measure the deformed length of each of these tube-like shelly fossils (called belemnites), and then make a close estimate of the original length of the shells by adding together the lengths of each of the dark fragments. Notice that the amount of extension of each belemnite is different, and depends on its orientation. Many features in rocks, such as fossils, mudcracks, or a stretched igneous dyke, may allow rough estimates of extension to be made.

Figure 8.4 Three deformed fossil belemnites. The darker fragments are part of the shells of the fossils, the white areas are where calcite crystals have grown in between the fragments.

8.3.2 Changes in angle during deformation

Angular changes are quantified by reference to two lines that were originally perpendicular in the undeformed rock.

Figure 8.5 shows an original square containing a circle deformed into a rhomb containing a strain ellipse. Deformation has taken place by the process of *shearing* (similar to pushing over a pack of cards). Each angle has changed by a certain amount known as the angle of **shear**, ψ (the Greek letter *psi,* pronounced 'sigh'), which can be calculated from any deformed angle *provided the original angle is known*. For example, in Figure 8.5b the angle α can be subtracted from 90° to calculate ψ, because the angle between the base and side of the square was originally 90° before deformation to a rhomb. The value of **shear strain**, γ (the Greek letter *gamma*) is given by:

$$\gamma = \tan \psi \qquad (8.3)$$

Question 8.2

In Figure 8.5b, what is (a) the amount of shear strain, γ, of the deformed square, (b) the percentage extension of the long axis of the resultant strain ellipse, and (c) the percentage extension of its short axis?
(*Hint*: as in Question 8.1, calculate extensions by comparing pre-deformation with post-deformation lengths.)

You have seen that extension can be difficult to quantify in real rocks, because the original *size* of a deformed feature is rarely known. Similarly, shear is hard to quantify unless information is available about the original *shape* of

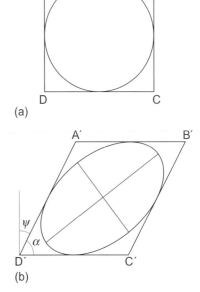

Figure 8.5 Distortion by shearing of a square block containing a circle (a) into a rhomb containing a strain ellipse (b). Note that the areas of the undeformed square ABCD and the deformed rhomb A′B′C′D′ are the same.

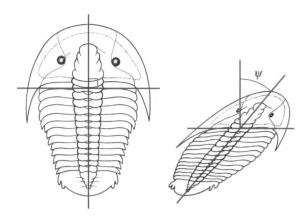

Figure 8.6 Undeformed and deformed fossil trilobites of the same species.

a deformed object (e.g. the angle between two undeformed lines). However, shear strain can be calculated wherever objects of known shape – particularly those containing natural right angles – have been deformed, like the fossil trilobites in Figure 8.6. Similar to extension, the amount of shear is generally different in different orientations.

One further point is very important. The ellipse formed by shear in Figure 8.5 is a similar elliptical *shape* to the ellipse shown in Figure 8.3; the main difference is its *orientation*. An elliptical shape in real deformed rocks (without the accompanying deformed square), could have been formed from a circle by squashing, or by shearing followed by rotation (or, for that matter, by squashing followed by rotation). This highlights the main practical difficulty in linking strain to strain-causing processes:

There is rarely enough information in a single deformed feature to identify – or quantify – the processes that deformed it, unless the shape and/or size of the feature before deformation is known.

8.3.3   Changes in area during deformation

The two deformations you have so far considered have not actually involved a change in area, but you can appreciate that such a change could happen, and is likely to happen in nature. In Figure 8.7, the short axis of the ellipse has shortened, but the ellipse long axis is the same length as the diameter of the original circle. The area of this ellipse must therefore be less than the area of the original circle.

Question 8.3

What is the axial ratio of the strain ellipse in Figure 8.7b?

Area change adds yet another complication to strain analysis. It is generally impossible to tell what part, if any, area change has played in deformation simply by measuring deformed shapes. The ellipse in Figure 8.7b is very similar in shape to both the ellipse in Figure 8.3b and to that in Figure 8.5b, yet each of the three ellipses formed by a different process. You can no doubt see that if the amount of deformation in each process was just right, the three strain ellipses would have *exactly* the same shape.

Although not many strain measurements yield precise, quantitative information about processes, they do provide very useful data with which to investigate processes.

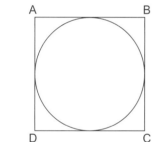

(a)

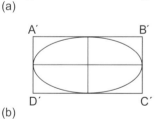

(b)

Figure 8.7 Distortion of a square block containing a circle (a) into a rectangle containing a strain ellipse (b) by area reduction. Note that the areas of the undeformed square ABCD and the deformed rectangle A′B′C′D′ are *not* the same.

8.4   Progressive deformation

So far you have considered deformation as a single event, which may occur as a combination of squashing, shearing and/or volume change. Yet there is another component: *time*. In nature, many rocks deform by the superimposition of successive increments of strain, rather than in just one single event.

The transition from circle to strain ellipse can be viewed as the superimposition of a series of small strains one upon another. Such deformation, generated through successive increments of shape change, is called **progressive deformation**. Each increment can be represented by a not-quite-circular strain ellipse. These ellipses may be superimposed *coaxially*, where the ellipse axes always lie in the same orientation (Figure 8.8a), resulting in **irrotational strain**. Alternatively, the strain ellipses may be superimposed *non-coaxially*, where successive ellipse axes lie in different orientations (Figure 8.8b). This is **rotational strain**. Three-dimensional, progressive, rotational strain is the usual case in naturally deforming rocks. During progressive deformation, lines in some orientations get longer whilst lines in other orientations get shorter. In some orientations, lines start off shortening and end up lengthening. Some lines never change length.

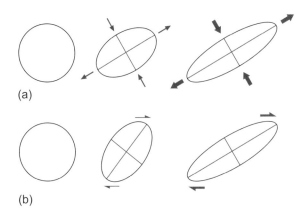

Figure 8.8   Diagrams showing progressive irrotational and rotational strain. In (a), small strain increments have been superimposed coaxially, whilst in (b), successive increments of shear have caused small strain increments to be superimposed non-coaxially. Note that in terms of shape and orientation, the two final ellipses in (a) and (b) are identical, even though they formed in different ways.

Activity 8.1   Lines with no extension

In this activity, you will explore some special features of extension in strain ellipses.

Activity 8.2   Progressive shear

In this activity, you will investigate length and angle changes during progressive shearing.

8.5   The main features of strain

How a line or plane in a rock deforms depends not only on what process caused the deformation but also on how that line (or plane) was oriented in the rock prior to deformation. During even a relatively straightforward process like shearing rocks with uniform properties, lines in different orientations extend by different amounts. Angles between lines in some orientations change rapidly whilst angles between others change slowly.

When rocks deform naturally, they rarely do so in a straightforward, tidy manner. Usually, both irrotational and rotational strain increments are involved. Commonly, the rock will also lose or gain volume during deformation. In Chapter 10 you will look at structures on a much broader scale, for instance in continental collision zones where horizontal compression of the Earth's crust dominates. In detail, however, such zones contain many types of structures formed by compression, extension and shearing in different orientations. Beds may be shortened in some directions, but lengthened in others, while reverse, strike-slip and even normal faults can occur in different parts of the collision zone. All these various structural features should be considered together if you want to describe the regional tectonic situation.

The most critical points to remember about deformation are summarised below:

- Whenever rock shortens in one direction, it *must at the same time* lengthen in at least one other direction, provided volume is conserved (which is not always the case). The converse is also true: lengthenings demand shortenings.

- Rotated features, such as dipping beds, do not automatically imply rotational strain. Lines and planes in most orientations rotate during irrotational strain as well.

- In general, most natural distortions are formed by a combination of non-coaxial strain (i.e. rotational strain) and volume change.

- Most natural strains can be considered as three-dimensional combinations of distortion, translation and rotation.

- Acute angles and strongly stretched linear features generally imply more strain, for example tight folds show greater strain than open folds.

8.6 Factors influencing strain in rocks

How rocks respond to stress depends on many physical variables, including:

- the *temperature* and *pressure* at which the rock deformed
- the *rate* at which the rock is deformed (strain rate)
- the *properties of the rock material* itself.

The physical properties of deforming rocks also change as the rocks deform. This might be because new minerals grow, or because beds change in thickness, or in particular because the nature of the fluids occupying the spaces within the rock changes. There are many possible variables – too many to consider in detail here. We shall concentrate on two broad categories: 'external properties' imposed on rocks from outside, for example temperature; and the 'internal properties' of the rock body itself, in particular its composition.

Activity 8.3 (optional)   Deforming everyday materials

This activity will give you a feel for the way strain is influenced by the physical properties of everyday materials.

8.6.1   Strain and the effects of pressure, temperature and time

The confining *pressure* at which a body of rock deforms depends on how deeply it is buried during deformation – the greater the depth, the greater the mass of the rock overburden and the higher the confining pressure. The *temperature* at which rocks deform depends on both the depth of burial *and*

the geothermal gradient. Hot rocks generally behave in a ductile way, cool
rocks in a brittle way (the same is true of chocolate). Rocks deformed at low
temperatures and pressures at or near the surface of the Earth (e.g. Figure 8.9a)
show far more evidence of brittle failure than those deformed at depth (e.g.
Figure 8.9b).

(a)

(b)

Figure 8.9   (a) A fault breccia with slate fragments in a finer orange matrix,
St David's, Wales. (Note penknife for scale.) (b) Ductile folds in a gneiss formed at
depth in the crust, Bhutan.

■   What structures should be common in rocks that have deformed near the
Earth's surface?

☐   Joints and faults, since they are the common brittle structures.

■   Why would major faults be expected to pass into shear zones at depth?

☐   The higher pressures and temperatures that exist at depth favour ductile
rather than brittle deformation, and shear zones can be considered as ductile
equivalents of faults.

The *rate* at which rocks deform is called the **strain rate**. The amount of change
over a given period of time is measured as strain per second. Since strain is
dimensionless, strain rate has units of s^{-1} ('per second').

Measurable deformation in rocks occurs over millions of years, not seconds. For
example, the Alps represent a tectonic event in which the crust shortened to about
half its original length and approximately doubled its thickness in 10–20 Ma. So
strain rates are always very small numbers.

■   What range of strain rates does the Alps event represent?

☐   Assume the Alps record a doubling in thickness, from an original thickness d
to a new thickness of $2d$. Using Equation 8.1, this is an extension of:

$$\frac{2d - d}{d} = \frac{d}{d} = 1 \text{ in } 10\text{--}20 \text{ Ma}$$

The strain rate is therefore between 1 in 10 Ma and 1 in 20 Ma. There are about 3×10^{13} seconds in 1 Ma, so in conventional units (s^{-1}), the strain rate ranges between

$$\frac{1}{10\,(3 \times 10^{13})\,s}, \text{ which is approximately } 3 \times 10^{-15}\ s^{-1} \text{ (to 1 significant figure)}$$

and

$$\frac{1}{20\,(3 \times 10^{13})\,s}, \text{ which is approximately } 2 \times 10^{-15}\ s^{-1} \text{ (to 1 significant figure).}$$

Such strain rates, roughly $10^{-14}\ s^{-1}$ to $10^{-15}\ s^{-1}$, are typical for naturally deformed rocks. $10^{-16}\ s^{-1}$ is slow, whilst $10^{-13}\ s^{-1}$ is fast. Under certain circumstances, natural strain rates can be much faster. For example, research has shown that magma movements are causing the summit region of Etna to deform at a staggeringly fast rate of $10^{-10}\ s^{-1}$.

Question 8.4

If the Alps had shortened at this extreme strain rate ($10^{-10}\ s^{-1}$), how long would the mountain range have taken to double in thickness?

Generally speaking, rocks deformed at high natural strain rates show more brittle behaviour than those deformed at low strain rates. Some brands of toffee show this effect. If you hit toffee with a hammer it will break, but you can also bend a toffee bar as long as you do it slowly. The brittle fracturing evident in Figure 8.9a reflects high strain rates during faulting; in some cases, this can generate so much frictional heat that rocks adjacent to the fault actually melt (Section 6.3.1).

External controls act in combination. Rocks deformed quickly at shallow depths usually deform in a brittle way, whereas rocks deformed slowly at greater depths usually deform in a ductile way. If you warm the toffee and flex it slowly, it should always bend, never break.

8.6.2   Strain and rock composition

If you spent some time looking at deformed rocks in the field, you would surely see a wide variation in structures – folds of varying wavelengths, amplitudes and shapes, some beds cleaved whilst others are not, faults here but not there, and so on. In any compact geographical region, it is almost inconceivable that the rocks have experienced widely different temperatures, pressures and strain rates during deformation. Variations in structures on a bed-by-bed basis (Figure 8.10) cannot wholly be explained by changes in external physical controls. The solution to this puzzle lies in the physical properties of the rocks themselves.

Rock *composition* has a profound effect on deformation. Some rock types are almost always rigid and tend to resist deformation; others almost always flow during deformation. This attribute is termed **competence**. Competent lithologies are

Figure 8.10   A calcareous mylonite from the Swiss Alps (field of view is about 0.5 m across). Most of the exposure is composed of marble, and has deformed in a ductile manner, producing a smooth, continuous foliation. However, some layers of a different composition were more competent, and deformed by brittle fracture, presumably under identical external conditions.

stiffer, flow less easily and break more readily than incompetent lithologies at the same temperature and pressure. Sandstones, limestones and most igneous rocks tend to deform in a competent way during deformation, whilst mudstones, shales and evaporites generally behave in an incompetent way.

■ In a deforming succession of interbedded sandstones and shales, would you expect faults or joints to initiate in the sandstones or in the shales? Why?

☐ In the sandstones. This is because, given similar pressure, temperature and strain rate, faults or joints (brittle failure) are more likely to form first in competent sandstones rather than incompetent shales (recall the distribution of joints in Figure 7.2a).

Structures are controlled by subtle variations in the rock successions, such as different thicknesses from layer to layer, changes in thickness of individual layers, particularly the thicker ones, or changes in spacing between layers. The presence and composition of pore fluids also have a marked effect on how rocks deform. The successions of rocks in different places are never absolutely identical, so the outcome of deforming rock successions can never be the same. Furthermore, the physical features discussed so far are only *some* of the factors that determine what types of structure develop.

One way forward from here would be to look at the simplest or most uniform of rock lithologies, and examine how variations in both physical parameters and strain components influence which structures form, then extend the investigation by degrees to more and more complex interlayers of rocks. However, you would quickly find that way to be lengthy, extremely complicated and generally unsatisfactory. Instead, Chapter 9 investigates the processes that lead to the formation of some specific, common structures. You will only get an insight into rock deformation and not an exhaustive account, but it will illustrate effectively how structures form.

8.7   Summary of Chapter 8

1   Stress is defined as force per unit area, and is measured in pascals. Stresses acting equally in all directions are known as pressure. The deeper you go within the Earth, the greater the confining pressure.

2   When stresses that are not equal in all directions act upon a body, they tend to deform it.

3   Strain is produced by stress. It can be defined as a proportional change in length (or angle, area or volume), relative to the original length (or angle, area or volume) and is therefore a dimensionless quantity.

4   Elastic strain is a recoverable form of ductile behaviour. Most rock structures are non-reversible, and therefore not elastic.

5   There is no simple mathematical link between stress and strain for real deformations.

6   Rocks may show strain as translation, rotation or distortion. Most deformations are a combination of all three.

7   During deformation, there are generally simultaneous changes in both lengths and angles. Changes in length are called extensions; changes in angle are called shears. A circle will deform into an ellipse – the strain ellipse – which reflects the amount and orientation of strain.

8   Progressive deformation is generated through successive increments of shape change. Small strain increments might be superimposed coaxially as irrotational strain, or non-coaxially as rotational strain. Progressive rotational strain is the usual case in naturally deforming rocks.

9   Whenever rock shortens in one direction, it *must* lengthen in another direction at the same time, if volume is to be preserved (though volume may change during deformation).

10   Rotated lines or planes do not automatically imply non-coaxial shearing. Most lines and planes also rotate during irrotational progressive deformation.

11   Acute angles and strongly stretched linear features generally imply greater strain.

12   Deformed features in rocks rarely allow the deformation process to be deduced or quantified, unless the shape or size of the feature before deformation is known.

13   How rock responds to stress depends on the temperature, pressure and rate at which it is deformed, the properties of the rock material itself, and many other variables.

14   Rock deformed near the surface of the Earth shows more evidence of brittle failure than that deformed at depth.

15   Rock deformed at high strain rates shows more brittle behaviour than that deformed at low strain rates.

16   Rock composition has a profound effect on deformation. Competent lithologies are stiffer, flow less easily and break more readily than incompetent lithologies.

17   Sandstones, limestones and most igneous rocks usually deform in a competent manner, whilst mudstones, shales and evaporites generally deform in an incompetent manner.

8.8   Objectives for Chapter 8

Now you have completed this chapter, you should be able to:

8.1 Distinguish between the concepts of stress and strain, and understand the difference between recoverable and permanent strains.

8.2 Explain the relationships between strain ellipses, extension and shear.

8.3 Use lengthened or shortened linear features in rocks to quantify extension, and changed right angles to determine shear.

8.4 Explain how natural strains can be considered as amalgamations of three-dimensional strain increments.

Now try the following questions to test your understanding of Chapter 8.

Question 8.5

Figure 8.6 shows two trilobite specimens, one deformed and the other undeformed. Determine the shear strain, γ. Why can you not measure the extension of the deformed trilobite?

Question 8.6

Study the deformed rocks in Figure 8.11, and then complete the description of the deformed features by inserting one of the missing terms into each of the spaces in the text. The missing terms are:

distorted   translated   folding   rotated   fracturing

'These siltstones initially deformed in a ductile manner, by _____, before further strain resulted in brittle _____. The hanging wall of the thrust has been _____ from right to left. The steep limbs of the folded beds have been _____ from their original horizontal orientation. The fact that once-planar beds are now curved and folded shows that they have been _____.'

Figure 8.11   A fold pair with an associated, sub-horizontal thrust fault at Broad Haven, southwest Wales (looking east; cliff in centre of view is about 20 m high).

Chapter 9   How structures form

This chapter examines in more detail how the structures outlined in Chapter 7 form under the influence of factors such as stress, temperature, pressure, strain rate and rock composition.

9.1   How brittle structures form

9.1.1   Stress and fracture orientation

The joints in Figure 7.2a show fractures that formed almost at right angles to the bedding planes. Do faults also form at right angles to bedding, like many joints? To answer this question, we must delve into stress in rather more detail.

What happens when real rocks are being deformed, if σ_{max}, σ_{int} and σ_{min} are all compressive and if σ_{max} is very much more compressive than σ_{min} (Figure 9.1a)? At first the rock shortens elastically parallel to σ_{max} and lengthens elastically parallel to σ_{min}, until it may fracture. Theory, supported by experiments, dictates that under compressive stresses a fracture will *always* develop at an acute angle to σ_{max} (angle α in Figure 9.1a). The angle α is always less than 45°, typically in the range 20°–40°. The fracture plane strikes parallel to σ_{int}. Although only one orientation of fracture plane is shown in Figure 9.1a, there is a mirror-image fracture that is just as likely to form. (Imagine viewing the page from behind, or hold a mirror up in front of the page to see how it looks.) Often fractures form in both possible orientations *at the same time*. When two such fractures form, they meet in an X-like pattern on surfaces perpendicular to σ_{int} (similar to the pattern of mineral-filled fractures in Figure 7.3). They are then described as *conjugate* fractures, as long as they formed at the same time.

The key to understanding the formation of faults lies in recognising that the presence of layers within deforming rocks actually controls the *orientation* of the stress field. The layers might typically be beds in sedimentary rocks or, perhaps, the surface of an igneous pluton, or a metamorphic schistosity, and so on. Usually one of σ_{max}, σ_{int} and σ_{min} is forced to act at right angles to the layering, whilst the other two act parallel to it.

Consider the case where sedimentary rocks near the Earth's surface are just starting to deform, and beds are still almost horizontal. Assume that, in this case, σ_{max} is the principal stress that is oriented at right angles to bedding (so σ_{int} and σ_{min} must both lie parallel to bedding).

- ■ In this case, is the rock body trying to lengthen or shorten along its layers?

- ☐ Since σ_{max}, the maximum compressive stress, is perpendicular to layering, the stresses acting along the layering (σ_{int} and σ_{min}) must be comparatively tensile and be acting to lengthen the layers.

- ■ If beds are trying to lengthen, and deformation is brittle, what sort of structure would you expect to develop?

- ☐ Normal faults should develop.

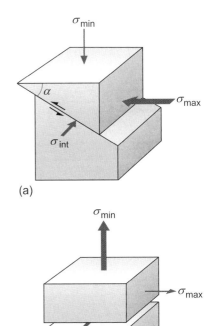

(a)

(b)

Figure 9.1   The development of fractures can be related to the magnitude of the principal stresses. (a) Under experimental conditions, failure under a *compressive* stress field (i.e. all principal stresses are compressive (positive), and $\sigma_{min} < \sigma_{int} < \sigma_{max}$) causes a fracture plane to develop, which lies at an acute angle α to σ_{max}. Angle α is typically 20°–40°. (b) Failure under a *tensile* stress field produces a fracture plane perpendicular to σ_{min} (all principal stresses are tensile, and hence *negative*). The largest magnitude tensile stress is the *most negative*, least compressive, stress, and so is designated σ_{min}. The relationship $\sigma_{min} < \sigma_{int} < \sigma_{max}$ therefore still holds.

Since failure planes always make a small angle with σ_{max}, normal faults should initiate as planes dipping steeply through the sedimentary layers (Figure 9.2a). One or both of the conjugate fractures may develop – if conjugate normal faults develop, a downthrown block or graben will form.

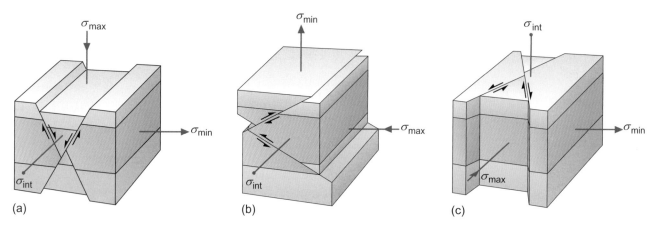

Figure 9.2   Three main types of fault that can initiate in layered rocks in response to different orientations of principal stress. (a) Conjugate normal faults form when the maximum principal compressive stress σ_{max} is perpendicular to layering. (b) Conjugate thrusts form when σ_{max} lies along layering and σ_{min} is perpendicular to layering. (c) Conjugate strike-slip faults form when both σ_{max} and σ_{min} lie parallel to layering.

Now imagine a situation where σ_{min} is perpendicular to bedding. Here, you would expect beds to shorten horizontally and thicken vertically. If brittle deformation occurs, you would expect reverse faults to develop. Since failure planes always make a small angle with σ_{max}, reverse faults should initiate as planes dipping gently through the sedimentary layers. Thrust faults (i.e. low-angle reverse faults) should therefore form (Figure 9.2b). These theoretical predictions fit very well with observed normal and reverse fault orientations, especially in deformed sedimentary rocks.

To complete the analysis, conjugate strike-slip faults should initiate when the principal stresses lie in a third specified orientation (Figure 9.2c).

9.1.2   Joint orientation

Some joints are associated with volume changes within the rock, for example the polygonal columnar joints seen in some lava flows and sills (e.g. Figure 4.5), which contracted as they cooled, or in the hexagonal cracks seen on the top of dried mud at the roadside. Joints occur in every rock type found on the surface of the Earth, so they are clearly not just dependent on lithology. As brittle structures they belong to the upper, colder parts of the crust – we have no evidence for the formation of joints deep down. What evidence is there of a link between the uplift and erosion (i.e. exhumation) of rocks and joint formation?

Some joints, especially in folded rocks, *do* have the same orientation as faults (even forming conjugate sets, as in Figure 7.3), but the majority of joints formed *at right angles* to bedding, in beds not obviously folded (e.g. Figure 7.2a). We know from experiments that if *all three* principal stresses are actually tensile, rather than compressive, rocks fail in the manner shown in Figure 9.1b. Fractures form perpendicular to the greatest tensile stress (i.e. the least compressive stress, σ_{min}), opening up a fracture in the rock known as an **extension fracture**. Bedding-perpendicular joints therefore suggest failure under conditions where the principal stresses were all tensile, and where σ_{min} lay parallel to bedding (Figure 9.3).

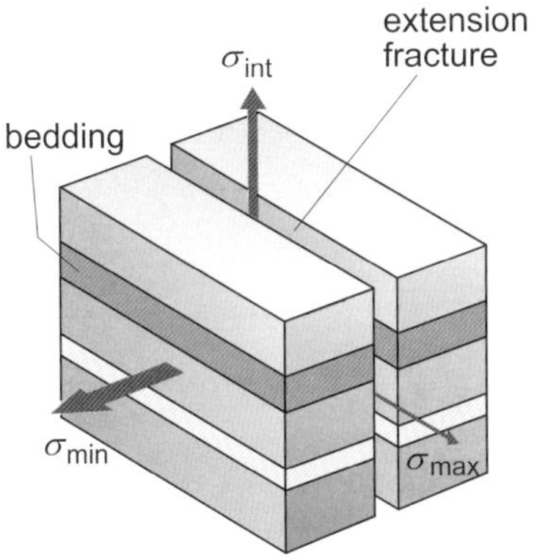

How can this come about, when tectonic stresses and the stresses due to rock overburden are compressive? Rocks respond to burial and tectonic stresses by deforming, but they also store some of those stresses elastically. When the rocks are brought back towards the surface, these stored stresses can exceed the minimal stresses due to the remaining overburden. The release of these stresses at shallow crustal levels causes extension fractures (i.e. joints) to form, often at right angles to rock layers as in Figure 9.3.

Figure 9.3 Development of an extension fracture under a tensile stress field (as in Figure 9.1b), but with σ_{min} parallel to bedding (shown as colour bands).

9.1.3 Boudins and competence contrast

Materials with a completely uniform composition, such as Plasticine, deform in the simplest way. What would happen if we deformed a block of Plasticine? We would not expect joints, faults or even folds to develop – the Plasticine block would simply shorten in some directions and lengthen in others, provided it is not so cold as to make it brittle.

Sedimentary rocks, however, consist of different beds with different compositions and thicknesses, and therefore different competences. When sedimentary rocks are strained, the effects of **competence contrast** can be seen. Suppose a layered composite of materials of different competences is stretched parallel to their length (Figure 9.4a). The competent layers may fracture, and separate into lozenge-shaped

(a)

(b)

(c)

Figure 9.4 Development of boudins. (a) Diagrammatic cross-sections through boudins in materials with a competence contrast (competence of materials: W > X > Y > Z; Z is the matrix). (b) Three stages through a laboratory experiment with layers of different competence in plastic modelling materials. (c) A real example of a boudinaged leucogranite layer, from northwest India. Lens cap for scale. (Leucogranite is a particularly pale-coloured granite, a feature reflecting the almost total lack of mafic minerals.)

pieces called **boudins** (pronounced 'boo-dans'), from the highly descriptive French term for a type of sausage.

Where the competence contrast is high, as it is between layers W and Z in Figure 9.4a, rectangular boudins separated by fractures form in the most competent layer (see Figure 8.10). Where the competence contrast is low, as it is between layers Y and Z in Figure 9.4a, the more competent layer flows and thins in places, and can even stretch without breaking. The theoretical situation shown in Figure 9.4a can be duplicated in laboratory experiments (Figure 9.4b), and is also seen in real rocks (Figure 9.4c).

9.1.4 Fault profiles

Faults are rarely as straight and planar as they appear on maps, because of the highly variable physical properties of rock layers. In particular, they typically show variations in dip, which are conspicuous when they are viewed in profile (i.e. in a section perpendicular to strike). Figure 9.5 is a series of profile views showing how faults often initiate and develop, using thrusts as an example.

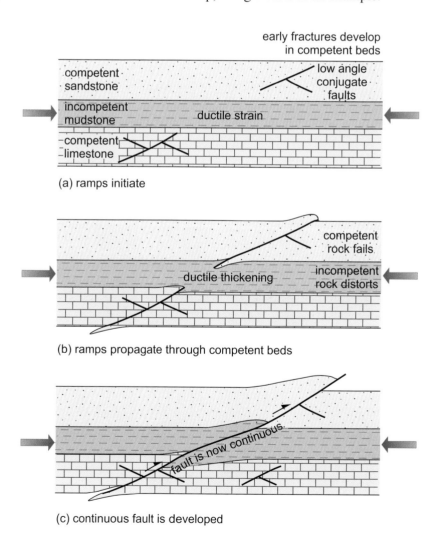

Figure 9.5 Stages in the initiation and development of a thrust fault in mixed sedimentary layers. The red arrows indicate that shortening is parallel to the layering. See text for explanation.

In Figure 9.5, a succession of sub-horizontal, undeformed sediments is shortened parallel to its layers. Fractures initiate in competent beds of limestone and sandstone at acute angles to bedding in accordance with the theory discussed above (Figure 9.5a). Either one of a pair of conjugate fractures could develop; in this example, leftward-dipping fractures grow. The incompetent mudstone that lies between competent beds must also be shortening at the same time, most probably by volume loss through expulsion of pore fluids from between the grains. As shortening continues, the newly formed fractures propagate right through the competent beds (Figure 9.5b), and eventually join along failure planes in the less-competent mudstone (Figure 9.5c). The thrust becomes continuous, allowing the hanging wall to move over the footwall.

The characteristic profile of a thrust is therefore a stepped one consisting of **flats** and **ramps**. Flats are those sections of the thrust plane that lie parallel, or almost parallel, to beds, whilst ramps are those sections that cut up through beds (Figure 9.6). Invariably, flats occur in incompetent horizons and ramps in more competent horizons. Most thrusts ramp upwards through their footwalls in the direction of transport of their hanging walls. Note that flats and ramps are defined relative to the orientations of beds, not necessarily to the horizontal.

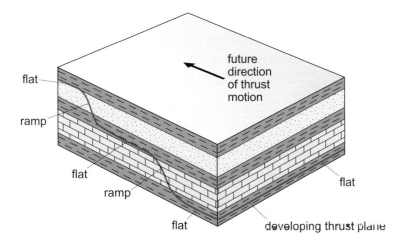

Figure 9.6 Flats and ramps developed in beds of contrasting competences, leading to a stepped fault profile.

Activity 9.1 Duplexes and other structures

In this activity, you will watch the final part of the video sequence *Structural Geology without Tears*.

9.2 How folds form

Folds form in two fundamentally different ways: as a mechanical consequence of fault displacement and as a mechanical consequence of shortening along layers.

9.2.1   Fault-bend folds

Activity 9.2   Flats, ramps and folds

Surprising things happen when the hanging wall of a stepped thrust moves over its footwall. Discover them now by doing Activity 9.2.

Imagine trying to push a rug up or down a set of shallow steps: it would crumple (develop folds) as it moved over the steps. The same is true of sheets of rocks moving on stepped faults – both anticlines and synclines may form as a hanging wall moves over a fault ramp. These are **fault-bend folds**, and an example of their geometry is shown in Figure 9.7. Although this example is for a thrust fault, stepped normal faults may also develop similar folds.

■   Look at Figure 9.7. How far down through the whole rock succession do fault-bend folds occur?

☐   The folds are restricted to the hanging wall block only. The footwall block has no fault-bend folds.

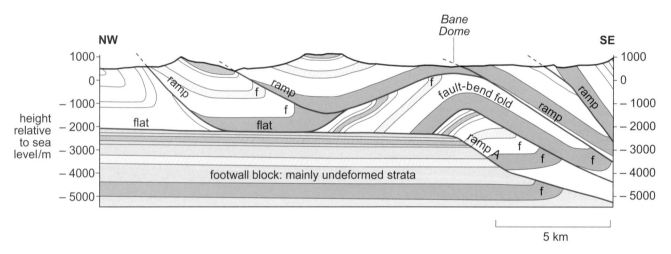

Figure 9.7   A major fault-bend fold developed above a ramp (ramp A) in a thrust fault. Several other minor folds (marked f) are tip-related folds (Section 9.2.2). This figure is a simplified cross-section through part of the Appalachian mountain belt in Virginia, USA.

9.2.2   Tip-related folds

Look again at Figure 8.11, which shows a thrust-related fold. First, you must make sure you can recognise the fault plane itself, which has a gentle apparent dip to the south (right). The fault plane shows up as a line that passes through the centre of the outcrop, below the anticlinally folded, pale sandstone beds. In the right-centre of the photograph, it marks a break between almost horizontal beds above and steeply dipping beds below. The prominent pale beds are displaced some 2–3 m to the left along this fault, so the hanging wall transport direction is from right to left (i.e. northwards).

■ Is this thrust folded?

☐ No.

■ Are beds in the footwall of this thrust folded?

☐ Yes.

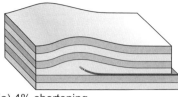

(a) 4% shortening

You can see several folds with the same sort of form in the cross-section in Figure 9.7 (marked f), including one in the otherwise undeformed footwall block. Clearly, some fold-forming process other than simple fault-bend folding is needed to explain why the footwall as well as the hanging wall has been deformed. This mechanism is illustrated in step form in Figure 9.8.

Immediately ahead of the tip of an advancing thrust lies an area that so far has not been deformed. As the 'envelope' of deformation approaches, strain increases in this area from zero through low strains towards intermediate strains. Strain rate also increases, and initial ductile strain (at low strain rates) would be expected to give way to brittle strain (at higher strain rates). Deformation will initially be achieved by ductile processes – folding and volume change rather than faulting. Figure 9.8a and b show a fold developing ahead of the advancing thrust tip.

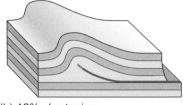

(b) 12% shortening

■ Is this fold pair (anticline and syncline) symmetric or asymmetric?

☐ Asymmetric. The leftward-dipping limb in Figure 9.8b is shorter and steeper than the rightward-dipping limbs (and eventually becomes overturned).

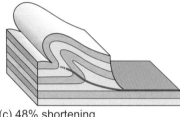

(c) 48% shortening

■ How does the fold asymmetry relate to the direction of thrust displacement?

☐ The geometry of the asymmetric anticline and syncline is consistent with the nature of the thrust displacement (i.e. the sense of fold asymmetry is towards the left, in the direction of thrust advance).

As brittle failure is reached (Figure 9.8c), the thrust propagates forwards through the newly formed fold. The upper, anticlinal part of the fold pair becomes part of the thrust hanging wall and is displaced away from its syncline, which is left behind in the footwall (Figure 9.8d).

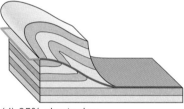

(d) 65% shortening

Figure 9.8 Stages in the development of a fold pair related to a thrust tip: (a) at first, ductile thickening ahead of the thrust tip is important, causing folds to form in the hanging wall; (b) asymmetric folds form, with a sense of overturning consistent with the direction of hanging wall movement; (c) the thrust propagates through the overturned limb of the fold; (d) finally, the anticline in the hanging wall is carried forward as the thrust tip advances beyond the fold.

Many folds initiate in front of thrust tip lines. They are typically asymmetrical, overturned in the direction of the thrust transport. Most folds that form during thrusting are a combination of two processes – fault-bend folding (Figure 9.7) and tip-related folding (Figure 9.8).

9.2.3   Buckle folds

Folds can also form as a mechanical consequence of shortening across layers, completely independent of fault movements. If a competent layer set in a less-competent matrix is shortened along its length, a mechanically unstable situation develops where the competent layer is deflected sideways and develops into a fold. These folds are called **buckle folds**, because they have formed by buckling – they have shortened along the length of the layers. They are generally

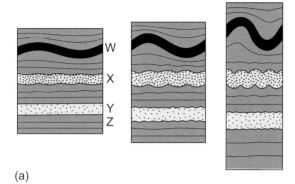

(a)

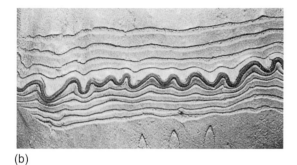

(b)

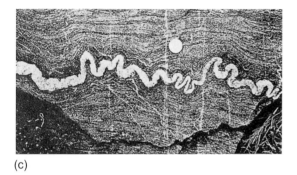

(c)

Figure 9.9 Development of buckle folds. (a) Diagrammatic cross-sections through buckle folds in materials with a competence contrast (competence of materials: W > X > Y > Z; Z is the matrix); (b) a laboratory experiment with layers of different competence in plastic modelling materials; (c) a real example from metamorphic rocks in Zimbabwe.

not directly related to faulting. You can make a buckle fold very simply by holding the two ends of a flexible plastic ruler and moving your hands towards each other.

Figure 9.9 illustrates the principles of buckle folding in a layered composite of materials of different competences shortening along its length. If competent layers are separated from other competent layers by a considerable thickness of incompetent material, the wavelength of the buckle fold is controlled by both the thickness of the buckled layer and the competence contrast between layers. An increase in the thickness of the competent layer or an increase in the competence contrast both lead to an increase in the wavelength of the fold.

Where the competence contrast is high, as it is between layers W and Z, buckle folds develop with long limbs and long wavelengths. Where the competence contrast is low, as it is between layers Y and Z, the folds have very short limbs. The idealised situation shown in Figure 9.9a can be duplicated in laboratory experiments (Figure 9.9b); it is also seen in real rocks (Figure 9.9c).

■ How does the orientation of the axial planes of the buckle folds in Figure 9.9b relate to the direction of shortening?

☐ The fold axial planes lie approximately at right angles to the original layering, and the shortening direction is parallel to the initial layering. So the folds are oriented with their axial planes approximately at right angles to the direction of shortening.

This question illustrates an important point about buckle folds; they are a ductile way of shortening a layered succession of rocks. In examples where it is clear that folds have formed by buckling, and where independent evidence exists to determine the overall shortening direction, folds are commonly oriented such that their axial planes lie roughly at right angles to the shortening direction.

9.2.4   Common fold profiles

Whether they are generated through fault-related processes or by buckling, we can consider folded layers in terms of two commonly seen fold profiles. Figure 9.10a shows a cross-section through **parallel folds**. In parallel folds, the layer keeps a constant orthogonal thickness right around the fold. This is possible only because the radius of the outer arc of the fold is significantly greater than the radius of the inner arc. Contrast this profile with that in Figure 9.10b, which shows a cross-section through **similar folds**. In similar folds, the shape of two neighbouring folded surfaces is almost identical. To achieve this, the orthogonal thickness of the folded layer has to vary; it is much thicker in the fold hinge than it is in the fold limb (especially for the dark layers in Figure 9.10b).

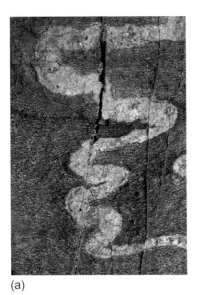

(a)   (b)

Figure 9.10   Fold shapes: (a) the folds in this granitic vein are mainly parallel, so the vein thickness does not change much around the fold hinges. Image 25 mm across; (b) successive folds in this layered rock have similar shapes (especially on the left-hand side): these are similar folds. Image 0.5 m across.

Parallel and similar folds are not extremes. You may see folds where the folded layer thins into the crest, or folds where thickening into the crest is so marked that the shape of adjacent folded layers is no longer identical. We highlight these particular fold shapes because natural folds are often a combination of parallel folds in competent layers and similar folds in incompetent layers. Parallel folds maintain bed thickness, but they cannot be stacked together indefinitely because of their rapidly changing arc radii. Similar folds can be stacked indefinitely, but require dramatic changes in bed thickness to develop. Natural folds are often a compromise between the two. In reality, the competence contrast between a buckling layer and its matrix is the most significant factor – high competence contrasts give rise to parallel folds whilst low competence contrasts produce similar folds.

9.3   How tectonic fabrics form

You have just seen that folds owe their existence to heterogeneities within the rock succession. Whether they form by buckling or by fault-related processes, competence contrasts between different lithologies must be present before folds can develop. What happens to uniform, homogeneous rocks as they deform?

9.3.1   Homogeneous lithologies and slaty cleavage

When a homogeneous rock, such as a mudstone, is deformed, it will shorten and thicken rather than generate folds or faults. How exactly does this happen? Mudstones are composed of many tiny grains of platy minerals and, depending on the exact rock composition, generally a percentage of minute quartz (or calcite) grains as well. Each of those individual grains must respond to deformation. Mineral grains deform in three ways: by distortion, by rotation coupled with translation, and by dissolution and regrowth. When grains are small, as they are in uniform mudstones, the last two are the more important processes.

Consider first what happens to platy grains such as micas and clay minerals during deformation. Most platy minerals are sheet silicates, composed of layers of silicate groups. Assume that the tiny mineral 'sandwiches' were originally randomly orientated in the undeformed rock (Figure 9.11a). As the rock shortens, almost all of these grains will rotate. Exceptions will be grains already oriented with their longest dimension either parallel to or perpendicular to the shortening orientation. As shortening (and attendant lengthening) of the rock progresses, platy grains will rotate first into a crude parallelism (Figure 9.11b) and eventually into an almost perfect alignment (Figure 9.11c).

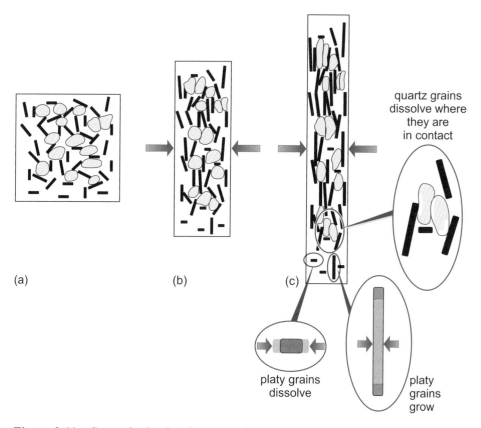

quartz grains dissolve where they are in contact

(a)   (b)   (c)

platy grains dissolve

platy grains grow

Figure 9.11   Stages in the development of a cleavage by shortening in mudrocks through grain alignment and recrystallisation. The rod-like shapes represent platy minerals, the more equidimensional grains represent quartz. Arrows indicate shortening direction. See text for explanation.

At the same time as physical rotation is taking place, increasing temperatures and pressures during deformation are driving mineralogical changes. Sheet silicates dissolve and recrystallise by subtracting or adding material from the *edges* of their lattice plates, rather than by subtracting or adding more plates. Whether any individual lattice grows or dissolves depends on its orientation relative to the shortening direction. Grains that lie parallel to the shortening direction have their ends highly stressed, so they dissolve. The phenomenon is known as

pressure dissolution (commonly referred to as pressure solution). Conversely, grains lying with their long axes at right angles to the shortening direction are in an ideal orientation to grow and will grow at the expense of those lying parallel to the shortening direction (Figure 9.11c).

However, the effects of pressure dissolution on small quartz or calcite grains within the rock are usually much more significant. These minerals are relatively equidimensional, and tend to dissolve away at points where they are in contact with other minerals (Figure 9.11c). This process removes material from the quartz or calcite crystal into solution, reducing the size of the grain, and allowing nearby platy minerals to rotate still further. Quartz or calcite is re-precipitated from solution onto the low-stress areas of existing grains, or into intergranular sites as new grains, or as quartz or calcite veins within the rock (Figure 9.12).

Figure 9.12   Sub-horizontal white veins of calcite that cut across pressure solution cleavage (steep, dark grooves), which is axial planar to open folds in mudstone and siltstone beds, Dyfed, Wales. The veins may have formed during the same deformation event as the cleavage, providing sites for re-precipitation of material dissolved during cleavage formation. Photograph is about 4 m across.

The combined effect of these processes of rotation and dissolution elongates grains physically and chemically. The end-product is a rock with a very strong texture, known as a **tectonic fabric**; in this case an alignment of platy grains. Rocks with such a fabric split preferentially along the plane defined by the oriented plates; they have *slaty cleavage* (Figure 7.31). A cleavage that forms dominantly by the process of pressure dissolution, known as a **pressure solution cleavage**, is commonly composed of cleavage planes that are spaced apart (e.g. Figure 9.12), and may even develop into alternating bands of soluble (generally pale) and less-soluble (typically dark) minerals. Many analyses of strain in slates have shown that slaty cleavage forms perpendicular to the direction of overall maximum shortening, which can be in excess of 50%. Volume reduction through pressure dissolution is also an important process during slaty cleavage formation.

It is rare, however, that vast thicknesses of sedimentary rocks are completely uniform in composition. Even in predominantly muddy successions, there are usually some beds that are sandier. Lithological variation provides competence contrast. This means that in successions of, say, mudstones with interbedded sandy beds, cleavage formation will be accompanied by buckle folding. Fold style will be controlled by the thickness and competence contrast of the competent lithologies. Cleavage can form in all lithologies, but slaty cleavage is always better developed in the muddy, incompetent layers than it is in the sandier, competent layers (Figure 9.13a).

The cleavage often also changes its orientation from layer to layer (Figure 9.13b). In folded competent beds, rather than being exactly parallel to the fold axial planes, the cleavage typically converges *downwards* towards the axial plane of an anticline. In less-competent beds, the cleavage 'fan' converges *upwards* towards the axial plane of an anticline (Figure 9.13c). Differential orientation of the cleavage plane from bed to bed is known as **cleavage refraction**.

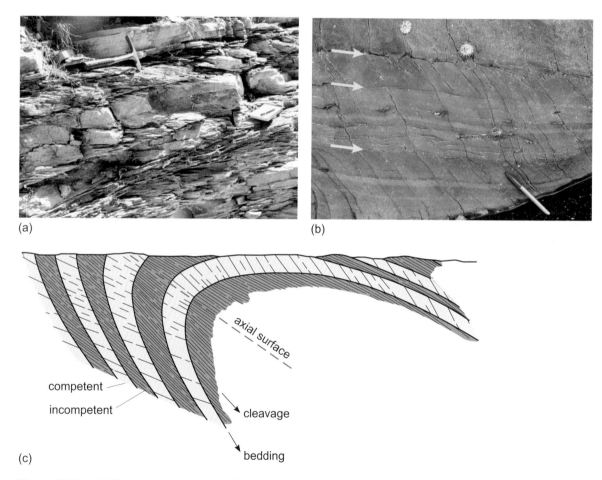

Figure 9.13 (a) Contrasting cleavage development in siltstones (strong cleavage) and quartzitic sandstones (cleavage weak or absent), NW India. (b) Refraction of spaced cleavage traces developed in mudstones and siltstones, west coast of Wales. Cleavage planes tend to form at higher angles to bedding in coarser sediments, such as the disrupted silty layer at the top of the photograph. Arrows mark bedding surfaces where pronounced refraction occurs. (c) Cleavage refraction in folded layers of different competence. Beige beds are competent sandstones; blue–grey beds are incompetent mudstones. The slaty cleavage orientation is shown by broken lines that are nearly parallel to the axial plane (or axial surface) of the fold.

It is important to realise that folds and cleavage are *independent* ways of achieving shortening in rocks. The axial planes of folds and cleavage planes both lie at right angles to the maximum shortening direction. Many folds carry an axial planar cleavage, but not all folds do. There are also places where the rock is cleaved, but not folded.

9.3.2   Metamorphic fabrics

So far, we have considered how tectonic fabrics develop mostly in sedimentary rocks, but more strongly metamorphosed rocks also almost always carry tectonic fabrics. Many metamorphic rocks have very long, complex histories. Not all the minerals present in the rock necessarily formed at the same time. Yet by looking closely at the texture of a metamorphic rock, to see the relative ages and orientations of its metamorphic minerals, we can usually unravel the history of changing pressure and temperature conditions, and relate these to the rock's deformation history.

Deformed platy minerals such as micas tend to stack alongside each other to form a planar fabric, or foliation (Figure 9.14a), whereas elongate minerals like amphibole, or stretched grains, tend to line up together, forming a linear fabric, or lineation (Figure 9.14b).

The presence of minerals showing a foliation or lineation in a metamorphic rock shows that the rock must have been deformed either during or after mineral growth. Several different mechanisms can account for the alignment of minerals in a metamorphic rock, yet each is a means by which the minerals, and therefore the rock itself, can be shortened.

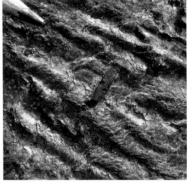

(a)

First, deformation can occur *within* individual grains through the movement of defects in the atomic lattice. The result is the elongation of the grain shape at right angles to the shortening direction. Lattice distortions caused by deformation are common in quartz, and can be recognised by wavy extinction when the rock is viewed in thin section between crossed polars (Figure 9.15). With further strain, the defects become tangled up together, with the result that the original grains become divided into much smaller grains known as subgrains.

(b)

Figure 9.14   (a) Crumpled foliation in a mica–schist, Scotland (note tip of pencil (top left) for scale). The edges of individual folia of aligned muscovite mica flakes can be seen on some parts of the fold hinges. (b) Aligned amphiboles forming a strong linear fabric that runs through this deformed amphibolite from Connemara, Ireland. The lack of planar fabric (foliation) is shown by the equigranular appearance of the right-hand face of the sample (sample is 86 mm across). (c) A high-strain gneiss from southern Spain with a planar foliation (parallel to the top surface) and a lineation oriented at 90° to the thin black line marked on the foliation surface (sample is 11 cm across). The rock has been flattened vertically to produce the foliation and, at the same time, stretched parallel to the lineation. Grains on the right-hand face, perpendicular to the lineation, appear merely flattened; grains on the left-hand face, parallel to the lineation, show the effects of both vertical flattening and horizontal stretching.

(c)

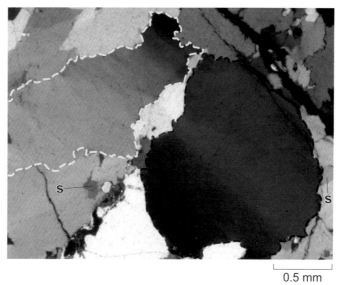

0.5 mm

Figure 9.15   Deformed quartz grains (outlined in red and yellow) showing wavy extinction in a rock viewed between crossed polars. Two subgrains are marked with '**S**'.

Secondly, deformation during metamorphism usually causes crystallisation and growth of *new* minerals in a preferred orientation. This can occur through nucleation of crystals with a common alignment, because crystals require less energy to nucleate and grow in certain orientations with respect to the shortening direction. For example, platy minerals tend to grow with the flat sides of their crystal faces at right angles to shortening. The rock develops a cleavage and elongate minerals grow with their long axes in the cleavage plane.

However, deformation does not always occur at exactly the same time as metamorphism. In fact, it is usually possible to determine the relationship between the time of deformation and the time of metamorphism by examining the fabric of the rock. For example, if metamorphism occurred before deformation, then early-formed metamorphic minerals will themselves be deformed. This is clearly shown by the distorted mica flakes seen in Figure 9.16.

Metamorphism must have occurred early in a sequence of deformation events if new minerals are aligned to form a planar schistosity, and a later episode of deformation affects this schistosity (Figure 9.17). The result is a shortening of the layers and fold formation. In Figure 9.17, the rock has been shortened along the horizontal axis of the photograph.

(a)

(b)

Figure 9.16   (a) Muscovite flake buckled by deformation of schist (plane-polarised light; width of image 2.2 mm). (b) The same grain seen between crossed polars.

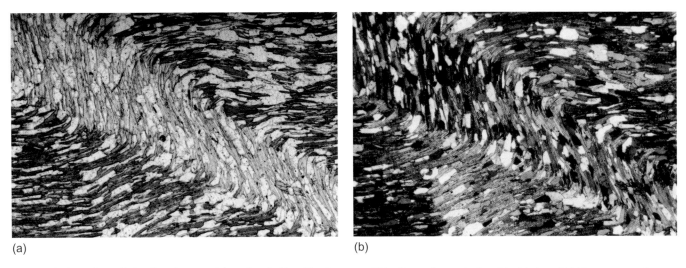

(a)   (b)

Figure 9.17   An originally planar schistosity defined by layers of biotite and quartz, buckled into a fold by a second deformation episode: (a) plane-polarised light, width of image 3.5 mm; (b) seen between crossed polars.

Where metamorphism occurs after deformation, new metamorphic minerals will not be aligned, but will instead cut across the fabric of the rock. Figure 9.18 shows well-formed crystals of andalusite (grey) that have grown in random orientations over a pre-existing schistosity, indicating that temperatures remained elevated after deformation ceased, promoting andalusite growth. Similar 'post-tectonic' mineral growth often happens during contact metamorphism, when hot magma intrudes previously deformed rocks, causing new minerals to grow randomly over the pre-existing tectonic fabric.

0.5 mm

Figure 9.18   Porphyroblasts of andalusite (straight-edged, dark and light grey) that have overgrown a schistosity defined by mica in a regionally metamorphosed schist (between crossed polars).

Most commonly, however, metamorphism and deformation occur at much the same time. The relationship between them can sometimes be evaluated by examining trails of inclusions within metamorphic *porphyroblasts*. For example, a garnet growing in a rock which already has a planar schistosity (Figure 9.19a) will grow around inclusions that lie parallel to the schistosity. Subsequent deformation (Figure 9.19b and c) may cause the porphyroblast to rotate and, if it continues to grow, the garnet will engulf inclusions around its rim at an increasing angle to the inclusions in its core (Figure 9.19d). The final result is an S-shaped or sigmoidal pattern of inclusion trails characteristic of mineral growth during deformation (Figure 9.19e).

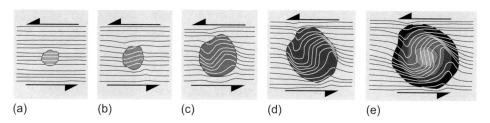

(a)   (b)   (c)   (d)   (e)

Figure 9.19   Sequence by which S-shaped inclusion trails develop in a growing garnet porphyroblast that is rotated during deformation. The four stages in the growth of the garnet (a–d) are shown in different shades within the final garnet crystal in (e), to highlight the rotation of the garnet core.

Activity 9.3   Interpreting metamorphic fabrics

In this activity, you will use a combination of simple rules and common sense to analyse fabrics in two different regionally metamorphosed rocks, as seen down a microscope.

9.4   Timing of rock deformation

To end this chapter, we touch briefly on how structural geologists date rock deformation. This information is necessary if we are to know how long any deformation episode lasted, or when an episode of deformation or metamorphism took place. The clearest indication of the age of deformation is given by the age of the rocks themselves. Rocks cannot deform before they came into existence, so if the original age of a deformed rock is known, then the *maximum* age of deformation can be established. Similarly, undeformed sedimentary rocks that lie unconformably on deformed rocks must be younger than the deformation, as must undeformed igneous rocks that are intruded into deformed rocks.

In recent years, it has become possible to perform radiometric dating on individual minerals such as mica or garnet. If it is clear from the rock fabric that the mineral grew before, during or after deformation, then dating the mineral puts precise limits on the age of deformation.

There are other ways in which the relative age of deformation can be established, albeit less precisely. It may be that both the limbs and axes of folds appear themselves to have been folded; these structures are often called **refolded folds**. The upright, tight-to-isoclinal fold in the lower-right corner of Figure 9.20 has been folded by a later upright open fold – if you sketch in the axial plane of the tight fold, you can confirm that it has been folded. When this happens, we can be sure that the folds *with folded axial planes* belong to a relatively early episode of deformation, which pre-dates at least one later folding episode. Although we know nothing of the exact age of either episode of folding (except that they are both younger than the rocks themselves), we do know the relative order in which the two fold episodes came.

Figure 9.20   Refolded fold. The upright tight-to-isoclinal fold in the lower-right corner has been folded by a later upright open fold. The area shown is about a metre across.

This type of observation can be extended to include folded cleavages. Folds of cleavage must indicate two distinct episodes of deformation – the first to form the cleavage and the second to generate folds in it (Figure 9.17).

In the last three chapters, you have seen that structures at all scales form as a result of tectonic stresses imposed on different rock types. You have learnt how to describe deformation features quantitatively using the concept of strain, and you have been introduced to the processes by which structures are formed, from a microscopic scale upwards to the scale of mountains. In Chapter 10, you will widen your view to look at structures on a larger scale, in the context of global tectonics.

9.5   Summary of Chapter 9

1   Normal faults form in response to horizontal stretching, often at high angles to sub-horizontal bedding. Reverse faults form in response to horizontal shortening, often at low angles to sub-horizontal bedding. Strike-slip faults allow the crust to change its lateral dimensions.

2   The orientation of rock layers often controls the orientation of the local stress field, and hence the orientation of faults.

3   Joints form by brittle failure; many joints are attributed to uplift and erosion.

4   The characteristic profile of faults, particularly thrusts, is one of flats and ramps, because many faults form from fractures that develop at different times in different layers. Flats lie parallel or almost parallel to layering, whilst ramps cut up through beds.

5   Fault-related folds form as a mechanical consequence of fault displacement. Buckle folds form as a mechanical consequence of shortening along layers.

6   In parallel folds, the layer thickness remains constant around the fold. In similar folds, the shape of two neighbouring surfaces is identical, but the bed is thicker in the fold hinge than in the fold limb.

7   When homogeneous lithologies deform, grains rotate physically, dissolve and regrow. The end-product is a rock with a fabric characterised by very strong alignment of platy grains, i.e. with slaty cleavage.

8   Cleavage and folding are independent ways of shortening a body of rock. However, both processes commonly act together, so that many folds have an associated, axial planar cleavage.

9   Several mechanisms account for the alignment of platy or elongate minerals in a deformed rock. Deformation can elongate grains through the movement of dislocations in the atomic lattice. Nucleation of new minerals with a common orientation may result in the rock developing a cleavage. Recrystallisation of minerals that do not have a platy shape, particularly quartz, can result in the breakdown of grains into subgrains.

10   Where metamorphism occurs after deformation, new minerals often overgrow the fabric of the rock.

11   A clear indication of the age of deformation is given by the age of the rocks themselves. If the original age of deformed rock is known, then that defines the earliest possible time of deformation. Undeformed rock must post-date deformation. Refolded folds and folded cleavage indicate two or more distinct episodes of deformation.

9.6   Objectives for Chapter 9

Now you have completed this chapter, you should be able to:

9.1 List the physical parameters that control whether rock deforms in a brittle or a ductile way.

9.2 Identify competent and incompetent lithologies from their behaviour within structures.

9.3 Explain how joints form, and why normal faults are usually steep whilst thrusts are usually low-angle structures.

9.4 Describe the separate processes by which folds form.

9.5 Explain why initially homogeneous rocks can develop a slaty cleavage.

9.6 Describe the mechanisms by which simple metamorphic fabrics are formed.

9.7 Interpret simple metamorphic textures as seen in exposures, hand specimens and thin sections in terms of the relative timing of deformation and metamorphism.

9.8 Determine the relative age of rock structures from geological evidence.

Now try the following questions to test your understanding of Chapter 9.

Question 9.1

Which of these structures – normal faults, ductile shear zones, thrust faults, joints – would you expect to form in the upper few kilometres of the crust, and which in the deeper crust?

Question 9.2

Suggest a way in which (a) a homogeneous, and (b) a heterogeneous body of rock could shorten in a ductile fashion. Which structures would characterise each of these processes?

Question 9.3

Identify the main differences between parallel and similar folds.

Chapter 10   Tectonic environments

The processes that occur at the boundaries of lithospheric plates – continental breakup, separation and continental collision – dictate the structural style of most deformed sedimentary and metamorphic rocks. In this chapter, we consider three tectonic associations that typify continental tectonics.

Firstly, there are areas of continental lithosphere characterised by thick successions of sedimentary rocks, relatively simple structures dominated by normal faults, subsidence over time, and a conspicuous lack of metamorphic rocks. These areas are called **sedimentary basins** and the dominant tectonic style is crustal extension.

Secondly, there are belts containing tectonically interleaved sedimentary and metamorphic rocks, and igneous (particularly granitic) intrusions. These belts are characterised by uplift, and by complex structures such as major thrust faults and large-scale folds with cleavages, which result from the shortening and thickening of crustal rocks. Typically, such belts form mountain chains – or at least the eroded remnants of former mountain chains. They are known as **orogenic belts**, from the Greek *oros* meaning 'mountain' and *genesis* meaning 'production'. Orogenic belts are often the result of continental collision.

Thirdly, there are smaller yet significant areas of continental crust that are not necessarily mountainous and may contain sedimentary, igneous or metamorphic rock, but which are characterised by large-scale lateral rather than vertical movements. These are known as **strike-slip zones**, which are often associated with transform margins.

10.1   Structures associated with crustal extension

We begin this section by considering extension within sedimentary basins, involving only the upper layers of the continental crust.

Sedimentary basins are characterised by thick successions of sedimentary rocks deformed by arrays of predominantly normal faults. Recall that normal faults are caused by stretching of the crust parallel to the Earth's surface, and thick accumulations of sediment can only be accommodated where the surface is actively subsiding. These two features suggest that basins form by regional surface-parallel extension accompanied by thinning normal to the Earth's surface.

◼ What structures would you expect to see near the surface in sedimentary basins?

☐ Near the surface, brittle structures would dominate because of the low temperatures. Normal faults would be common, and horsts and grabens may form. Sedimentary successions may be extensively jointed.

Figure 10.1 shows a structural cross-section through part of the North Sea basin. In the centre of the basin, up to 8000 m of sedimentary rocks have been deposited in the 250 Ma since Permian times, cut by large normal faults throwing down towards the Viking Graben, which lies in the centre of the basin. These large normal faults separate tilted blocks some 10 km wide.

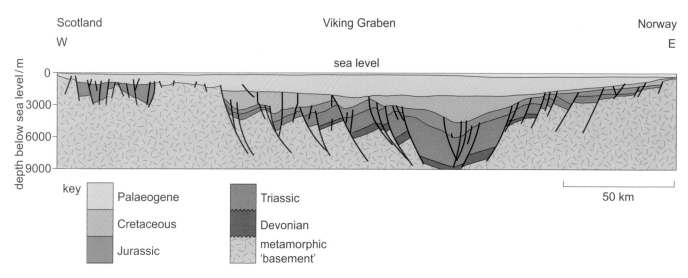

Figure 10.1   A cross-section through the upper crustal layers of the northern North Sea sedimentary basin. Note the scale of this basin, spanning more than 250 km between Scotland and Norway. Heavy black lines are faults. Vertical scale is ×5 greater than the horizontal scale.

■   What would be the implications, in terms of brittle and ductile structures, of surface-parallel extension deeper down, below the cross-section in Figure 10.1?

□   At depth, ductile behaviour would be expected, probably expressed both as ductile shear zones and ductile thinning.

Notice that the normal faults in Figure 10.1 are nearly all curved with dips decreasing with depth. Faults with this geometry are termed **listric** faults. Just how the North Sea basin formed is shown diagrammatically in Figure 10.2. When the lithosphere thins because of crustal stretching, the upper surface must subside. The base of the crust (marked by the Moho) rises, bringing the hot mantle nearer to the Earth's surface (Figure 10.2b). The results are an increase in geothermal gradient and hence heat flow through the stretched zone and an associated thermal expansion leading to isostatic disequilibrium.

When extension stops, faults become inactive. The upper layers of hot lithospheric mantle cool, and heat flow wanes (Figure 10.2c). As the layers cool, they become denser and sink, pulling down the surface of the stretched area. This space will become filled with sediments. The basin will therefore have two structural components: a lower one containing faulted and rotated blocks of older sediment that represents the *extension* phase, and an upper one containing unfaulted, younger sediment that represents a *thermal collapse* phase. This process has happened in the North Sea, as you can see by comparing Figure 10.1 with Figure 10.2c.

■   Looking at Figure 10.1, by when had most of the faulting and block rotation taken place?

□   By Palaeogene times, because all but the oldest Palaeogene rocks are unaffected by faulting.

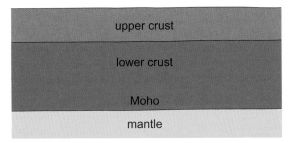

(a) prior to extension

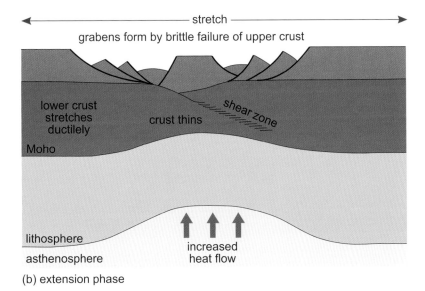

(b) extension phase

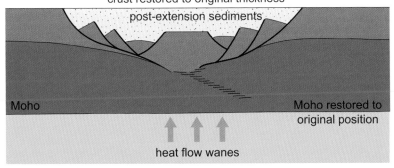

(c) thermal collapse phase

Figure 10.2 Formation of a sedimentary basin by extension followed by thermal collapse. (a) The upper part of the lithosphere prior to stretching. (b) The lithosphere thins as it stretches, and its upper part fails in a brittle manner, giving rise to rotated blocks separated by normal faults. Heat flow increases because hot mantle rocks are near the surface, and thermal expansion leads to isostatic uplift. (c) When stretching stops, heat flow wanes and the basin subsides, again due to isostatic disequilibrium. More sediment is deposited. Note the two structural components of the basin: an upper unfaulted part and a lower faulted part.

Continued crustal extension can lead to decompression melting of the rising mantle. In past tectonic cycles, there have been frequent examples of continents splitting along a zone of crustal extension, as shown in Figure 10.3. Extension is initiated (Figure 10.3a), and as separation continues, the rising lithospheric mantle starts to melt. Volcanic rocks become increasingly important within the basin (Figure 10.3b). If the rift is underlain by a mantle plume, a great volume of basaltic magmatism – flood basalts – may be generated (Section 5.4.3). The process ends in the separation of two bodies of continental crust and the formation of a new zone of oceanic crust between them. The Red Sea is a good example of continental crust under extension. Oceanic crust is beginning to form along its central zone as Arabia moves away from Africa (Figure 10.4). As each continental margin moves away from the new spreading axis, it cools and subsides. The continental margin becomes blanketed by sedimentary rocks. The edge of the continental crust is now a passive continental margin, and spreading has become focused at a mid-ocean ridge.

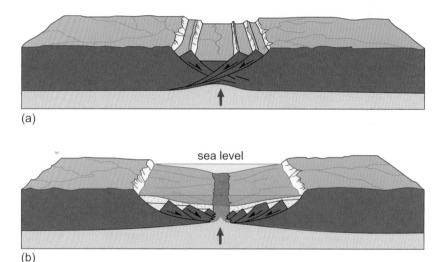

(a)

(b)

Figure 10.3   Cross-sections of the sequence of events at a divergent continental plate boundary. At first (a), the upper parts of the crust extend by developing a series of brittle normal faults. The continental surface also sinks, creating a basin that provides a site for the accumulation of sedimentary or volcanic rocks. As the plates continue to diverge (b), the lower lithosphere will rise and melt. Oceanic crust forms in the centre of the basin and the continent has separated into two.

Figure 10.4 Satellite image of the Suez rift (foreground) widening to the linear Red Sea basin, looking south with Africa on the right and Arabia on the left.

10.2 Structures associated with continental collision

If sedimentary basins and passive margins form at the start of the continental plate tectonic cycle, how does the cycle end? Continents cannot continue to separate indefinitely.

■ Why not?

☐ Because the globe is finite in size. New oceanic lithosphere created at a divergent plate boundary must be compensated for by subduction at a convergent plate boundary.

Separating continental fragments, each trailing a passive margin, are at the same time moving towards other continental crustal fragments. Eventually the continents must meet.

We have a rough idea of the timescale over which this happens. Except for ophiolites – those rare parts of oceans that have become 'fossilised' often within the suture zones that mark former plate boundaries (Section 5.4.5) – the oldest

Figure 10.5   The central Nepalese Himalaya (Gauri Shankar, 7134 m) seen from the south, the result of continental collision about 50 million years ago followed by uplift as an isostatic response to crustal thickening by thrusting and folding.

oceanic crust on the globe today is about 200 Ma old. All older oceanic crust has been subducted at convergent plate boundaries. The implication is that, after about 200 Ma, passive continental margins must convert into active, destructive plate boundaries. Since it took about 200 Ma to start and form an ocean basin fully, then oceans should close and continents should converge within roughly 400 Ma. Plate tectonics is believed to have been operating on Earth for at least 2500 Ma – long enough for many cycles of separation of a megacontinent into fragments, leading to ocean formation and continental drift, leading to continental collision and megacontinent reconfiguration.

Continental collision causes deformation, crustal thickening and consequent uplift from isostasy. The result is some of the most spectacular topography on the planet, for example the Himalaya (Figure 10.5).

When the Indian continental fragment separated from a giant southern continent about 120 million years ago, it journeyed a quarter of the way around the globe before colliding with Asia about 50 million years ago. A host of Himalayan structural evidence and metamorphic rock ages of between 40 Ma and the present time suggest that there is a strong correlation between the impact of India into Asia and the formation of the Himalayan mountain belt.

Colliding continents form orogenic belts with complex structures. Since passive continental margins have a strongly extensional form, any continental margin approaching a subduction zone is also likely to have an extensional form. Hence any structures that form during collision will be superimposed onto pre-existing extensional structures (Figure 10.6a).

As two continents meet and impact, one of them has to ride up over the other. This action will introduce a regional-scale shear to the deforming zone (Figure 10.6b). Which block overrides the other is controlled by the direction in which the subduction zone is dipping; the continent attached to the subducting oceanic slab tends to be pulled underneath the approaching continent. Any regional shearing component will control the asymmetry of structures.

We can now consider the scale and timing of the deformation (Figure 10.6c). Initially, only rocks at the oceanward margin of the continent are deformed. As collision proceeds, deformation spreads outwards and downwards away from the impact zone towards the interiors of both continents. Over time, more and more continental crust becomes incorporated into the orogenic belt. Deformation towards the continental interior will in general be slightly younger than deformation in the original impact zone. New thrusts develop in the outermost part of the belt, carrying deformed rocks away from the collision site and towards the continental interiors. Deformation becomes 'frozen' at its maximum extent, when continental collision finally runs out of energy and stops.

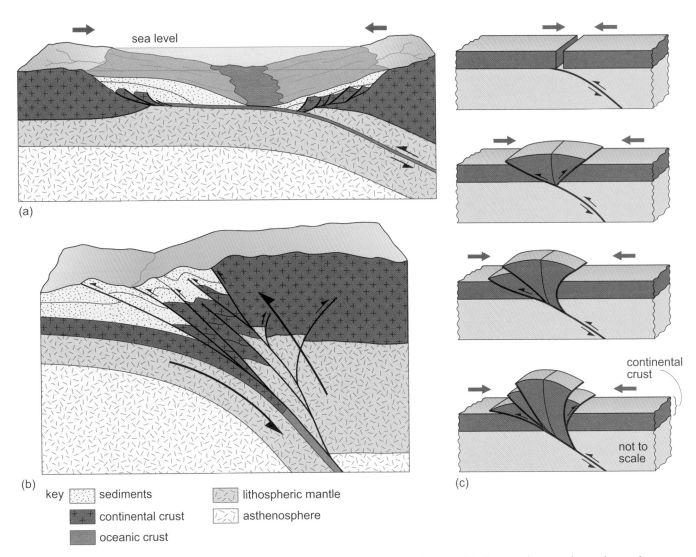

(a)

(b)

(c)

continental crust

not to scale

key ⬚ sediments   ▨ lithospheric mantle

☩ continental crust   ▨ asthenosphere

▬ oceanic crust

Figure 10.6   Three key characteristics of continental collision made in the text. (a) Converging continental margins already have an extensional geometry. (b) Orogenic belts are asymmetric because one continent must override another. The sense of shear (large, black half-arrows) will typically be dictated by the direction of subduction. (c) Time sequence showing that the margins of both continents are the first to be involved in deformation. At successive stages, the deformation 'front' moves outwards into the continental block. Sequentially younger thrusts develop in the outer parts of the orogenic belt, carrying deformed rocks in their hanging walls. The belt as a whole shortens and thickens.

Within the orogenic belt, there is an outer zone and an inner zone. In the outer zone, rocks from the continental interior dominate (e.g. thin layers of sedimentary rocks that rest unconformably on much older metamorphic rocks), and the effects of increased pressure and temperature are less. Deformation is relatively modest, and dominated by thrusts and fault-related fold structures in sediments or low-grade metamorphic rocks. Generally, they have deformed over a relatively short period of time, at modest temperatures and pressures. The only high-grade metamorphic rocks present are slices of the 'basement', typically gneisses and schists, on which the sedimentary rocks were deposited.

Structures that are typical of the outer parts of many orogenic belts are shown in Figure 10.7. Shortening has been achieved both by folding and thrusting. Thrusts show displacements ranging up to several kilometres, and thrust planes dip towards the inner part of the orogenic belt. Hanging wall displacement is usually away from the site of collision. Folds are commonly asymmetric, open, steeply inclined, and gently plunging. The rocks often carry only a weak, sporadically developed cleavage, and typically only one episode of folding has occurred. The metamorphic record indicates low pressures and temperatures during deformation. Granite plutons are rare or absent.

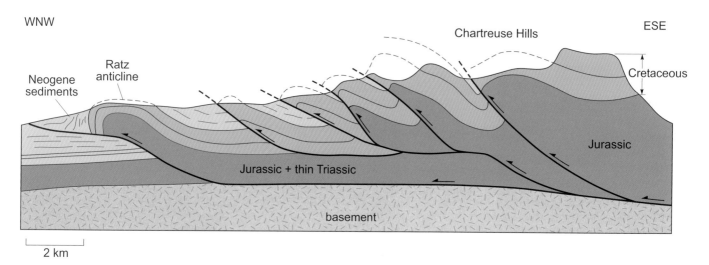

Figure 10.7   A true-scale cross-section through part of the outer zone of the Alps, near Grenoble in France. Heavy lines are thrusts.

Many of these features can be seen in Britain's best-preserved orogenic outer zone – the Moine Thrust Belt in the northwest Highlands of Scotland. The region is characterised by major thrusts that have emplaced older rocks, Precambrian gneisses, on top of younger rocks, Cambrian quartzites (Figure 10.8).

Figure 10.8   The Glencoul Thrust (Loch Glencoul), part of the Moine Thrust Belt, looking northeast. All the rocks below the red thrust plane are Cambrian quartzites. Above the fault are Precambrian gneisses (Lewisian), which have been thrust over the quartzites.

The outer zone of the Alpine orogen also exposes a sequence of major thrusts.

■ Looking at Figure 10.7, when did the folding and thrusting take place?

☐ In Neogene times, because Jurassic, Cretaceous and Neogene rocks are folded, and Jurassic/Cretaceous rocks are thrust over Neogene rocks.

The inner parts of orogenic belts typically show complex ductile structures in high-grade metamorphic rocks. Deformation is intense and complex; early thrusts and fault-related folds are later deformed by ductile folds and schistosities. Because of the increased pressures and temperatures, the rocks have experienced medium- to high-grade metamorphism with complex fabrics. Some lithologies have started to melt.

Virtually no two inner orogenic belts are alike. Each major orogen seen around the world tends to have its own distinctive organisation of its internal parts. That applies not only to young mountain belts, such as the Alps and the Himalaya, but also to the eroded remnants of very old mountain belts, such as the Scottish Grampian Highlands or the Limpopo Belt of southern Africa. However, several features seem to be common to the inner parts of many orogens.

First, the rocks involved are often much more varied in lithology than those in the outer parts of the orogen. This reflects the complex extensional history that the more oceanward margins of the continents experienced on their original separation.

Secondly, there are repeated episodes of deformation, usually all recording severe crustal shortening. Often many discrete 'phases' can be recognised by refolded folds or multiple cleavages. Early folds are commonly asymmetric, tight to isoclinal, flat-lying to recumbent, and usually have a strong axial planar cleavage. Later folds are usually also tight and may be either upright or inclined, and often carry an associated cleavage or schistosity.

Thirdly, the metamorphic record indicates high to extreme pressures, and medium to high temperatures during deformation. There is often evidence that highly deformed gneisses have begun to melt to produce veins of granite magma (Figure 10.9). These are preserved as migmatites.

Fourthly, brittle faults are much less common than ductile folds and cleavages. However, interior zones frequently show late normal faults that are believed to originate through the 'collapse' effects of gravity acting on the thickened crust.

The boundary between the outer and inner zones may be marked by a clear tectonic line, which is often an important thrust. The mechanisms by which plate movement causes tectonic structures are still hotly debated amongst Earth scientists. However, the general principles are agreed. Converging plates are driven primarily by gravity acting on the cold, dense slab of subducting oceanic crust. Continental crust cannot be subducted because of its relatively low density. Shear forces generated by the

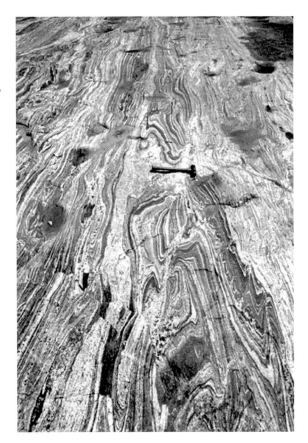

Figure 10.9 Intense ductile deformation of Precambrian biotite gneisses and migmatites with melts forming pale, granitic veins and layers, from Sand River, South Africa.

dipping slab act on both continental blocks (Figure 10.6b) and produce horizontal shortening and vertical thickening which continue as long as plate movement continues. Throughout deformation, the overthickened continental crust collapses under its own weight, helping to drive thrusting in the outer parts of the orogen, and generating significant internal normal faults.

Some of the world's mountain belts, for example the Urals in Russia, lie towards the centre of continents rather than at their margins. Such mountain chains belong to an earlier cycle of plate tectonics and lie at a join between two once-separate continents. The most important such join in Britain lies through northern England and Ireland from the Solway Firth to the Shannon estuary. This line is all that is left of a former ocean that once separated a continental block that included England, Wales and southeast Ireland from another block that included Scotland and northwest Ireland. The ocean that formerly separated these two continents, called the Iapetus Ocean, closed about 420 Ma ago (an appropriate name as, in Greek mythology, Iapetus was the father of Atlantis, and the Iapetus Ocean can be considered as the forerunner of the Atlantic Ocean). The attendant collision formed a major mountain belt, the remnants of which can still be seen in the igneous and folded metamorphic rocks of the Scottish Highlands, and the fold and slate belts of Southern Scotland, the Lake District and North Wales. This event is known as the Caledonian Orogeny.

10.3 Structures associated with transform margins

Transform fault boundaries (defined in Section 2.3.4) allow lithospheric plates to move past one another and are dominated by strike-slip movement. In these zones, lateral movement of blocks is the major way in which the crust is deforming; compression and extension play a minor role. Despite the dominant movement being horizontal, most structures that develop in strike-slip zones also have vertical displacements.

Natural strike-slip faults have many bends, and the structures that develop at these bends are shown diagrammatically in Figure 10.10. One type of bend allows small sedimentary basins to develop. In these zones, normal faults will form at a high angle to the major strike-slip faults, and open up small, rhomb-shaped basins (Figure 10.10a). The opposite type of bend generates structures that shorten and uplift the rocks near the bend. In these zones, shortening will be accommodated by both thrusting and folding, perhaps with cleavage development (Figure 10.10b). Again, the trace of the structures will be oriented at a high angle to the main strike-slip faults.

Major strike-slip faults typically split into branches or strands that join together further along the length of the fault. Where the main fault splits, or where two strands of the fault overlap, there are generally the same spatial problems accommodating lateral displacements as those seen at fault bends.

If major strike-slip faults act under a thick sedimentary cover, folds can develop that often show a distinctive pattern above the fault. You can easily get the idea by placing a cloth over two breadboards and then sliding one board sideways past the other; the cloth will ruck up into a set of folds. These folds will be arranged

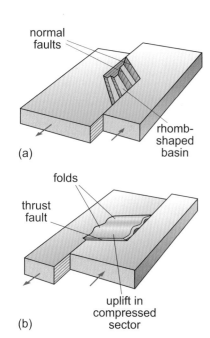

Figure 10.10 Structural features that develop in association with bends in strike-slip systems: (a) extensional features that develop where strike-slip movement opens up space; (b) compressional features that develop where strike-slip movement closes down space.

along the underlying fault, but in detail each will be an anticline that plunges in both directions away from its centre, and each will be slightly offset relative to the next (Figure 10.11).

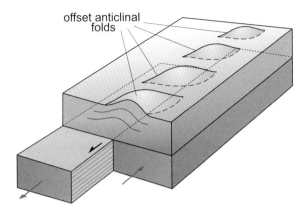

Figure 10.11   Offset anticlines forming in the sedimentary cover above a buried strike-slip fault.

Strike-slip deformation dominates along transform fault boundaries where plates are moving past, not towards or away from, each other. As we mentioned in Section 2.3.4, the San Andreas Fault in Western USA is perhaps the best-known continental transform plate boundary. Indeed, many of our formative ideas on strike-slip tectonics came from this plate boundary. Its sense of displacement is dextral, but the zone itself is not straight, so there are many instances of both sedimentary basins, and zones of fold-thrusts or offset anticlinal folds along its length. It is one of the zones in the world where folding and faulting are actively taking place on the surface and where strain can be measured.

One of the most intriguing aspects of the San Andreas Fault is that it is only midway through its plate tectonic life. When the Pacific Ocean closes, as it surely will at some time in the future, the structures along the San Andreas Fault will become overprinted by younger, probably more significant, structures associated with continental collision.

It is clear that strike-slip structures play an important role in deformation, from the metre scale of single faults to the hundreds-of-kilometres scale of transform fault boundaries. If blocks of crust are to move, they have to be detached from their adjacent crust. Strike-slip structures are one of the two important facilitators of this; the other is large-scale detachments. Consequently, we would expect to find strike-slip structures in each of the other tectonic environments, and at all scales. We would also expect these lines, once developed, to be reactivated. The Great Glen Fault in northern Scotland is a good example; it is a major strike-slip line that formed within an orogenic zone, yet was later reactivated as part of a basin-forming episode (Figure 10.12).

Figure 10.12   The Great Glen Fault, looking northeast. This is the most spectacular tectonic feature of the British Isles resulting from a major strike-slip fault, which extends across northern Scotland from the Atlantic to the Moray Firth. The fault runs southwest from Inverness (NH (28) 64) on the Bedrock UK Map) through Loch Ness to Loch Linnhe (NM (17) 95). Reproduced with the permission of the British Geological Survey © NERC. All rights Reserved.

Activity 10.1   The Great Glen Fault

This map-based activity evaluates the displacement on the Great Glen Fault.

Strike-slip movement on the Great Glen Fault over tens of kilometres opened up adjacent areas under extension allowing sedimentary basins to develop, into which the Old Red Sandstone strata were deposited. This movement has been largely dextral although periods of sinistral motion have also been deduced from geological evidence.

In this final chapter, we have glimpsed the complex structures that are found in common tectonic settings. We have established that extensional, compressional and strike-slip structures are all found in more than one tectonic setting. For example, collision zones often incorporate structures formed during the pre-collision history. Early work in the Himalaya erroneously accredited all structures and metamorphism to the well-documented Indo-Asian Eocene collision, yet recently more detailed work has unravelled evidence for pre-collision extension and even for a previous collision event in the Palaeozoic. Because continental crust cannot be subducted, it is important to remember that it is likely to preserve evidence for a long-lived geological evolution. Indeed, it is probable that every orogenic zone on all of the Earth's continents has seen more than one tectonic event.

10.4   Summary of Chapter 10

1   Irrespective of age and global location, certain types of structure repeatedly occur together. Extensional structures predominate at constructive plate boundaries and shortening structures predominate at collision zones.

2   Sedimentary basins are areas of continental crust that have thick deposits of sediments deformed by arrays of normal faults. They form by regional surface-parallel extension accompanied by contraction normal to the Earth's surface.

3   In many sedimentary basins, an early period of crustal stretching and faulting is succeeded by a later period of thermal collapse without significant faulting.

4   Most large sedimentary basins that developed in the past were either intracontinental graben systems, or passive margins that lay between oceanic and continental crust as a result of continental separation.

5   Continental collision gives rise to a wide and varied zone of deformation within an orogenic belt. As collision proceeds, deformation spreads outwards away from the impact zone towards the interior of the continent. Zones are established in which deformation towards the continental interior is generally slightly younger than deformation in the (formerly oceanic) impact zone. Deformation becomes 'frozen' at its maximum extent when continental collision ends.

6   The outer parts of orogenic belts typically show simple structures in sedimentary rocks or low-grade metamorphic rocks. Shortening is achieved both by folding and displacement on thrusts. Thrusts dip towards the inner

part of the orogenic belt, and commonly show hanging wall movement away from the site of collision. Folds are commonly asymmetric, open, steeply inclined and gently plunging. The rocks themselves usually carry only a weak cleavage. Often only one episode of folding has occurred.

7   The inner parts of orogenic belts typically show complex ductile structures in high-grade metamorphic rocks. They show repeated episodes of deformation. Early folds are commonly asymmetric, tight to isoclinal, flat-lying to recumbent, and usually have a strong axial planar cleavage. Later folds are usually also tight and may be either upright or inclined, and often carry an associated cleavage. Their metamorphic record indicates high pressures and temperatures during deformation.

8   Strike-slip structures play an important role in deformation, from metre scale to the scale of transform plate boundaries. Locally they may include areas of extension and areas of compression.

10.5   Objectives for Chapter 10

Now you have completed this chapter, you should be able to:

10.1   Describe the main tectonic associations on the Earth's continents.

10.2   List the main structural features by which you would recognise sedimentary basins, the outer and inner parts of orogenic belts, and strike-slip zones.

10.3   Outline how each tectonic association fits into the plate tectonic cycle.

Now try the following question to test your understanding of Chapter 10.

Question 10.1

From Figure 10.7, what is the % extension in this part of the outer Alpine orogenic belt: −10%, −30% or −50%? First take a guess, to see how good your 'feel' for the amount of deformation is, then make a more accurate calculation to check how good your guess was. (*Hint*: you need to know the length of any one bed before and after deformation. We recommend you use the bed in the middle of the Cretaceous rocks.)

Answers to questions

Question 2.1

The average density of the Earth is much greater than the density of surface rocks. One explanation is that the interior consists of a totally different type of material that is much denser than rock. Alternatively, pressure must increase within the Earth due to the weight of overlying material and you would therefore expect rock at depth to be compressed to greater densities. The Earth's interior would therefore be denser than the Earth's surface even if the interior and surface were made of material of identical composition.

Question 2.2

(a) The SiO_2 content of oceanic crust falls into the range of SiO_2 content of mafic igneous rocks given in Table 6.4 of Book 1. Its overall composition is a close match to the gabbro shown in Table 6.3 of Book 1.

(b) The SiO_2 content of continental crust falls into the range of SiO_2 content for intermediate igneous rocks given in Table 6.4 of Book 1.

Question 2.3

The time is equal to the distance divided by the speed at which the subducting plate is approaching the trench. First, make sure that the distance and speed use the same units for distance (e.g. metres). 80 km is 8×10^4 m, and 1 cm y^{-1} is 1×10^{-2} m y^{-1}. The time to move 80 km at this speed is:

$$\frac{8 \times 10^4 \text{ m}}{1 \times 10^{-2} \text{ m y}^{-1}} = 8 \times 10^6 \text{ years, or 8 million years.}$$

Question 2.4

(a) $\dfrac{\text{mass}}{\text{volume}} = \text{density}$

The fact that the Earth's average density is so much greater than the density of material near the surface tells you that there must be something unusually dense (now known to be the core) deep in the interior.

(b) Seismic studies reveal the compositional layers, in particular the boundaries between the inner and outer core and between the outer core and mantle, and the Moho between the crust and mantle. There are also jumps in seismic speeds attributed to phase transitions, such as the one between the upper and lower mantle.

(c) The Earth's magnetic field indicates a fluid and electrically conducting outer core.

(d) The observation that previously ice-covered areas are being uplifted is explained by isostatic adjustments following the removal of ice. These adjustments require that the interior of the Earth is behaving like a very viscous liquid.

Question 2.5

(a) Removal of material from the top of the crust means that the crust will become thinner. In order to maintain isostatic equilibrium, the base of the crust will rise. [*Comment*: As the base rises, it must push the whole column up with it, which will tend to encourage even more erosion at the top.]

(b) Deposition of sediment on top of column D will add an extra load, so that the whole column will be isostatically depressed.

(c) Flow takes place in the asthenosphere, which is below the bottom of Figure 2.13. The mantle within the diagram is part of the lithosphere and will bend (flex) to allow the crust to rise and fall, but it is too strong to flow.

Question 2.6

These are: (i) isostasy, as explored in Question 2.5c; (ii) plate tectonics, which can only happen because there is a weak layer below the lithosphere that allows the plates to move around.

Question 2.7

False. There is a divergent plate boundary (the Mid-Atlantic Ridge) running down the centre of the Atlantic (shown in Figure 2.4), therefore North America and the western part of the North Atlantic lie on a different plate to Europe and the eastern part of the North Atlantic. Similarly, South America and the western part of the South Atlantic lie on a different plate to Africa and the eastern part of the South Atlantic.

Question 2.8

Heat-producing elements are more abundant in continental crust than in oceanic crust, so the setting with the thickest continental crust is where most heat will be generated. This is (c), where the crust has been thickened by collision.

Question 3.1

(a) The graph shows a positive correlation between mean effusion rate and final lava flow length. However, there is scatter in the data and it falls into two groups: low effusion rate eruptions that produced short flows, and high effusion rate eruptions that produced long flows.

(b) An explanation could be that a fast-flowing lava (associated with a large effusion rate) will travel farther before becoming cold enough to stop advancing.

(c) If the effusion rate could be measured while the eruption is still going on, then the graph could be used to estimate the final length that a lava flow might reach. This could give local authorities guidance as to whether to evacuate people living and working downstream from the lava. This method is likely to be most reliable at the volcanoes used to plot Figure 3.8 because other volcanoes may have different magma properties or ground slopes that lead to a different relationship between mean effusion rate and final flow length.

Question 3.2

The umbrella cloud is the part at the top of the eruption column that spreads out. It occupies about half of the total height in the case of the eruption in Figure 3.14. The lower part of the convective ascent region and all of the gas thrust region are obscured.

Question 3.3

The contours of tephra thickness are displaced towards the south-southeast, indicating that the umbrella cloud was dispersed in that direction. The wind must therefore have been blowing from the north-northwest. If there had not been a strong wind blowing, then the isopachs and isopleths would have been arranged concentrically around the vent.

Question 3.4

Reading from the graph, a plume height of 12 km implies an eruption rate of a little more than 10^3 m^3 s^{-1}.

Question 3.5

(a) The conditions plot at point A on Figure 11.1 and falls in the stability field for a convecting column, so there would be a plinian eruption.

(b) The point representing a 400 m diameter vent and an exit velocity of 300 m s^{-1} (B on Figure 11.1) plots outside the field for a stable column, so the column would collapse, generating a pyroclastic flow.

(c) The new point (C on Figure 11.1) plots in the collapsing column field, so this decrease in water content would also lead to column collapse.

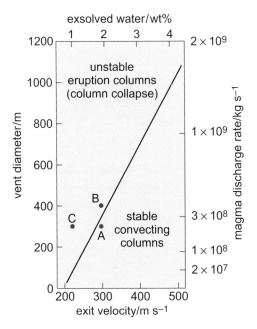

Figure 11.1   Annotated version of Figure 3.22, to accompany answer to Question 3.5.

Question 3.6

Because pyroclastic flows travel along the ground, we expect them to travel primarily down slopes and to be channelled along valleys. Flow deposits should therefore be thicker in the flat parts of valley bottoms (Figure 11.2a). In contrast, topography can have no influence on the transport of airborne tephra, the thickness of a fall deposit will be independent of topography but will reflect distance from the vent and direction of the wind (Figure 11.2b). Fall deposits mantle topography (Figure 3.16) whereas flow deposits fill topographic lows.

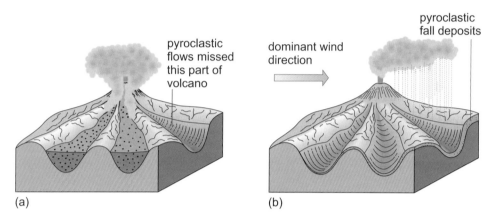

Figure 11.2   (a) Pyroclastic flow deposits are largely confined to valleys. (b) Fall deposits are distributed around a volcano independently of topography, but are controlled by wind direction. (To accompany the answer to Question 3.6.)

Question 3.7

The correct answers are shown in Table 11.1.

Table 11.1 Completed Table 3.2.

	Low volatile content	**High volatile content**
mafic magma	(c) pahoehoe and a'a lava flows	(b) fire fountains
felsic magma	(a) lava domes	(d) ignimbrites and pyroclastic fall deposits

Question 3.8

An ignimbrite usually contains pumice clasts together with lithic fragments in an ashy matrix. However, this is also the case for other types of pyroclastic deposits from felsic magmas. Ignimbrites have a distinctive threefold division into a ground layer (layer 1) overlain by the main deposit (layer 2) which is usually pumice-rich towards the top and lithic-rich towards the base. The uppermost layer (layer 3) is a thin layer of fine-grained ash. If the ignimbrite was of the welded variety, this would be recognised by the occurrence of squashed glassy pumices known as fiamme.

Question 4.1

The earthquakes were about 5 km from the caldera centre at 2200 hours on 10 July, and 28 km away at 1900 hours on 11 July. They therefore travelled about 23 km in about 21 hours. So the average speed was about 1.1 km h^{-1}, or 0.30 m s^{-1}. The gradient of the data in Figure 4.1d decreases over time, so the speed decreased over time. Up until about 1100 hours on 11 July, the speed was about 0.4 to 0.5 m s^{-1}, and after that time the speed was about 0.1 m s^{-1}.

Question 4.2

(i) Thin section (a) has a clearly visible groundmass of crystals (mostly plagioclase and pyroxene, but with some small opaque crystals that are likely to be iron oxide minerals) that are about 0.1 mm or less in size. Fine grain size is classified as <0.25 mm, so this rock is fine-grained. Most of thin section (b) has no discernible crystals, and consists of a dark (in fact brown) glass.

(ii) The glassy groundmass of thin section (b) shows that it comes from closer to the margin. In fact (b) is from within 1 cm of the margin, whereas (a) comes from several centimetres further in.

(iii) The relatively large plagioclase feldspar crystals must have begun to grow before the magma was emplaced (like the pyroxenes in the Whin Sill chilled margin in Figure 4.6). In (a), it is surrounded by fine-grained crystals that grew in the rapidly cooling magma as it lost heat to the country rocks; in (b), it is enclosed in glass resulting from the even more rapid chilling of the magma at the contact with the country rock. Because they are so much bigger than the crystals in the groundmass, these crystals can be described as phenocrysts (Book 1, Section 6.1) or, because they are rather small, as microphenocrysts.

Question 4.3

(a) During emplacement of cone sheets, block A is uplifted. This implies that the magma chamber is inflating, causing forceful injection of cone sheets through the updoming roof.

(b) During emplacement of a ring dyke, block A subsides along the associated fracture. Because of the outward dip of the fracture, this movement widens the fracture and magma must be displaced upwards to occupy the space. This is a more passive mode of emplacement than for cone sheets.

Question 4.4

The compositions of the four rocks are shown plotted on the classification diagram in Figure 11.3. The rocks are therefore classified as:

(a) syenite [Comment: Syenite is an alkali-rich felsic plutonic rock.]

(b) granodiorite; strictly speaking, because it is medium grained (rather than coarse grained), its correct name is micro-granodiorite.

(c) tonalite.

(d) gabbro and diorite, because of the low % of quartz and alkali feldspar. To decide whether it is dioritic or gabbroic, you have to take account of the composition of the plagioclase, which in this case is 70% albite. As this is more than 50%, the rock is diorite rather than gabbro.

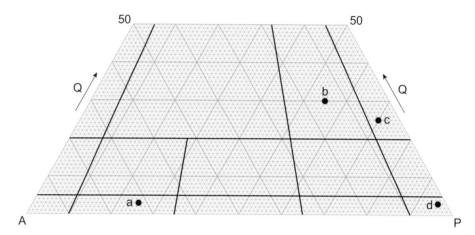

Figure 11.3   Classification diagram with the field boundaries of Figure 4.14, with the locations of points a to d given in Question 4.4.

Question 4.5

(a) This is a basaltic (mafic) dyke (i.e. a minor intrusion). Its composition is indicated by the dark-green colour, and its age (Palaeogene) by the letter G (italicised to show it is igneous). The fact that it is a dyke can be inferred from its narrow, near-linear outcrop discordant with the sedimentary units it crosses. The youngest of these is P1 (Lower Permian), so the dyke must be younger than the Lower Permian. (In fact this is one of the radial dykes from Mull.)

(b) This is a granite pluton, and is discordant with the country rock, cutting across the boundaries between D2 (Mid Devonian) and D3 (Late Devonian) and various Carboniferous strata (mainly C1–4 and C5–7). It must, therefore, be post-Westphalian (an epoch towards the top of the Carboniferous Period) in age, confirmed by the letters CP (Carboniferous and Permian) marking this unit. (In fact it is the Dartmoor granite and this pluton has a radiometric age of 280 Ma, i.e. it is Permian).

Question 4.6

The rock contains a total fraction of felsic minerals of 70%. Of this, the percentage of quartz is:

$$\frac{10}{70} \times 100\% = 14\%$$

and the percentage of feldspar is:

$$\frac{60}{70} \times 100\% = 86\%$$

With equal amounts of alkali and plagioclase feldspar, there is 43% of each. The rock lies in the centre of the quartz monzonite field of Figure 4.14.

Question 4.7

A completed Table 4.1 is shown below:

	Sill	Lava flow
chilled margin at top	✓	(✓)
chilled margin at bottom	✓	(✓)
columnar joints	✓	(✓)
concordant top with local discordance	✓	×
concordant base with local discordance	✓	(✓)
rubbly top	×	✓
rubbly bottom	×	✓
baking of overlying rock immediately above the contact	✓	×
baking of underlying rock immediately below the contact	✓	✓

Proper chilled margins occur in sills but not in lava flows. However, the glassy top of some kinds of pahoehoe might be mistaken for a chilled margin, which is why we have put a tick in brackets in the lava flow column. Similarly, the base of a lava flow might be chilled against the underlying rock, therefore chilled margins are not infallible guides. Columnar joints can occur in thick lava flows as well as in sills. Sills are generally concordant, but can be locally discordant at the top or the base. The base of a lava flow is discordant if there has been erosion prior to its eruption, but otherwise it is concordant. However, deposits laid on top of the flow will be concordant everywhere, with no local discordances. Rubbly tops and bottoms are characteristic of a'a flows and blocky flows, and offer the most reliable means in this table of distinguishing a lava flow from a sill. Sills cause baking of the rock into which they are intruded (Section 4.1) and similarly the heat from a lava would be expected to bake the immediately underlying rock. Material deposited on top of a lava flow cannot be baked, though, because the flow would be cold by then.

[*Comment*: A further criterion, not listed in the table, is weathering. The top of a lava flow may show the effects of long-term exposure prior to burial, whereas this cannot occur in a sill.]

Question 5.1

(a) The positions of the four points specified are plotted on Figure 11.4. Points (i) and (iii) lie between the solidus and the liquidus, so at these conditions mafic material would consist of a mixture of solid and liquid. Point (ii) lies to the left of the solidus and indicates that the sample is entirely solid. Point (iv) lies to the right of the liquidus, and indicates that the sample is entirely liquid.

(b) At 0.1 MPa, the solidus lies at a temperature of 1080 °C, so melting begins at this temperature. At the same pressure, the liquidus lies at 1200 °C, so the sample would become completely molten at this temperature.

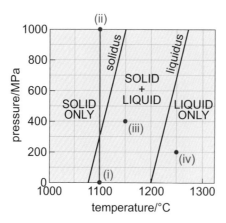

Figure 11.4   Annotated version of Figure 5.1. Points (i) to (iv) are referred to in the answer to Question 5.1. The line linking points (i) and (ii) is referred to in the answer to Question 5.2.

Question 5.2

Initially, the basalt will be entirely solid (the *P–T* conditions correspond to point (ii) on Figure 11.4). The *P–T* conditions experienced by the sample will vary along the straight line drawn from (ii) to (i) on Figure 11.4. Point (i) corresponds to the final state, which is a mixture of crystals and liquid. The first liquid will form under the conditions represented by the point where the line crosses the solidus, which is when *P* has decreased to about 300 MPa. The sample will be entirely solid until *P* has fallen to this value, and as *P* continues to decline the proportion of liquid will increase.

Question 5.3

The *P–T* conditions are those of point (ii) in Question 5.1a. Under anhydrous conditions this corresponds to completely solid basalt. However, if conditions become water-saturated, then the water-saturated phase boundaries apply. The water-saturated liquidus lies to the left of this point (Figure 5.2), so the sample would be completely molten (and would therefore be water-saturated mafic magma).

Question 5.4

Example (a) shows a very glassy rock with few tiny crystals, indicating very few nucleation sites and very low growth rates of those tiny crystals. This requires a relatively large undercooling (see Table 11.2). Example (b) has a few large crystals (phenocrysts), so nucleation rate was relatively low, but growth rate relatively high, suggesting low undercooling. Example (c) has a large number of small crystals, indicating high nucleation rate and low crystal growth rates.

Table 11.2 Completed Table 5.1.

Nucleation and growth rate	Undercooling	Example in Figure 5.5
Nucleation rate low, growth rate high	low	(b)
Nucleation rate high, growth rate low	moderate	(c)
Nucleation rate very low, growth rate very low	high	(a)

Question 5.5

Figure 5.6 shows that the temperatures mostly lie below the solidus of dry peridotite but above the solidus of wet peridotite. The implication is that partial melting will occur where the wedge has received water from the slab, and where the temperature and pressure lie above the solidus.

Question 5.6

Comparing Sections 5.4.1 and 5.4.4 (and Figure 5.14), the differences between magmatism at mid-ocean ridges and subduction zones can be summarised as follows.

		Mid-ocean ridge	Subduction zone
(a)	Depth and process of magma generation by partial melting	Partial melting of mantle peridotite by decompression melting of rising asthenosphere as it ascends below a divergent plate boundary. Partial melting starts at about 45 km depth	Partial melting of mantle peridotite by addition of water to the mantle wedge, released by dehydration reactions in the subducting slab. This causes the solidus of peridotite to fall, allowing partial melting to happen at temperatures that would otherwise be too low. Partial melting occurs between about 100 km (top of slab beneath volcanic arcs) and the local Moho
(b)	Depth and degree of fractional crystallisation and crustal assimilation	Fractional crystallisation of olivine, plagioclase and pyroxene produces evolved basalts at relatively shallow depths below mid-ocean ridges. No crustal assimilation because basaltic crust has too high a melting temperature to be assimilated by basalt	Fractional crystallisation of olivine, plagioclase, pyroxene and amphibole at various depths from deep crust to shallow sub-volcanic levels. Assimilation of intermediate and felsic crust. Possible partial melting of lower crust as a result of heat from intruded basalt
(c)	Composition of the erupted lavas	Basalt	Basalt, andesite and rhyolite

Question 5.7

(a) At 30 °C km^{-1}, temperature at a depth of 30 km would be higher than the surface temperature by 30 °C km^{-1} × 30 km = 900 °C. Surface temperature is 10 °C, so the temperature at 30 km depth is 910 °C.

(b) The point representing a depth of 30 km and a temperature of 910 °C is labelled A on Figure 11.5. This is to the left of the anhydrous solidus, showing that under these conditions anhydrous felsic material would be entirely solid.

(c) The water-saturated phase boundaries are now applicable. Point A lies to the right of the water-saturated liquidus, so the material will be completely molten.

(d) The line representing cooling at 5 °C km⁻¹ is plotted on Figure 11.5 (at 5 °C km⁻¹, cooling during a rise of 30 km would be 5 °C km⁻¹ × 30 km = 150 °C). The magma would be completely liquid until the depth where this cooling line crosses the liquidus (point B on Figure 11.5), which in this example is at 3.8 km depth (100 MPa). The magma will begin to crystallise at this depth, and become progressively more crystalline as it continues to rise (unless fractional crystallisation allowed the melt to be separated from the crystals, which we will ignore). The magma would be completely crystallised by the time its cooling path intersects the water-saturated solidus, at about 2.3 km (point C on Figure 11.5), and so it would cease to rise at this depth, if not slightly deeper. [*Comment*: This answer ignores the role of volatiles, which will tend to exsolve as the pressure declines.]

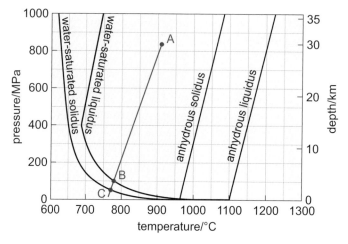

Figure 11.5   Annotated version of Figure 5.3, to accompany answer to Question 5.7.

Question 6.1

(a) The phases tremolite and calcite and quartz are all solids (minerals) and thus this assemblage is more ordered, and so has a lower entropy than the assemblage diopside + H_2O + CO_2, which contains two gases.

(b) The higher entropy assemblage, in this case diopside + H_2O + CO_2, is stable at higher grades of metamorphism.

Question 6.2

(a) At 300 MPa and 700 °C, a sample would lie in the sillimanite stability field (see A, Figure 11.6), thus andalusite would not be stable.

(b) At 800 MPa and 500 °C, a sample would lie in the kyanite field (B, Figure 11.6), so kyanite would be the stable phase.

(c) Kyanite and sillimanite only coexist in equilibrium on the phase boundary between their respective stability fields. At 700 MPa, the temperature on that phase boundary is about 660 °C (C, Figure 11.6).

(d) Kyanite is metastable, since at 400 °C and at atmospheric pressure andalusite must be the stable phase (D, Figure 11.6).

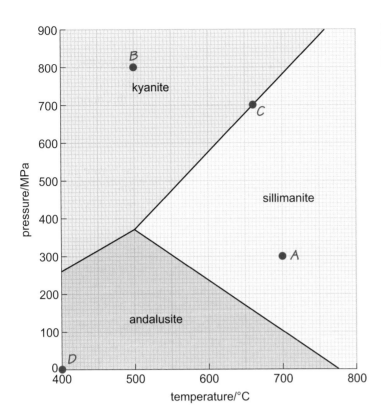

Figure 11.6 Phase diagram of the polymorphs kyanite, andalusite and sillimanite with points A, B, C and D marked for answers to Question 6.2.

Question 6.3

(a) Basalts contain feldspar, and although its composition may change during metamorphism, feldspar is often present in metamorphosed basalts. At high grades of metamorphism, garnet may form from the elements Fe, Mg, Ca and Al. However, there is insufficient K to form muscovite, and absence of carbon (in the form of carbonate ions) prevents the formation of calcite.

(b) Limestones are rich in Ca and CO_2, and thus calcite ($CaCO_3$) will be present after metamorphism. Conversely, limestones contain little or no aluminium, iron, magnesium or potassium and hence cannot form garnet (which requires Fe, Mg and Al) or muscovite (which requires K and Al) or feldspars (which need Al).

(c) Garnet, muscovite and feldspar occur in metamorphosed mudstones that contain Fe, Mg, Ca and Al. Absence of carbonate prevents calcite from forming.

Question 6.4

(a) Since andalusite rather than kyanite reacts to form sillimanite in the metamorphic aureole in Figure 6.12, the pressure must be less than about 370 MPa. Andalusite is not stable at higher pressures (Figure 6.4a).

(b) At the outer margin of the sillimanite zone in Figure 6.12, andalusite presumably reacted to form sillimanite. At a pressure of 200 MPa, this takes place at 625 °C according to the phase diagram in Figure 6.4a.

(c) Since 1000 MPa = 35 km (Section 6.1), then 200 MPa = 7 km. When the granodiorite was intruded, the thickness of overlying rock was 7 km.

Question 6.5

The reaction chlorite + quartz $\rightleftharpoons$ garnet + H$_2$O is temperature-dependent and almost independent of pressure at pressures greater than about 500 MPa; that is, the phase boundary is almost vertical on the phase diagram of pressure against temperature (Figure 6.16). At pressures of >370 MPa, garnet forms from quartz and chlorite at temperatures of 552 to 570 °C.

Question 6.6

(a) Only two isograds can be drawn on Figure 6.17: a garnet isograd between the biotite and the garnet zones, and a kyanite isograd between the garnet and kyanite zones (Figure 11.7).

(b) The isograds cut across the folds and so the metamorphism is younger than the folding.

(c) The isograds are not concentric around the intrusion, which occurs in an area of relatively low metamorphic grade (the biotite zone), and so the metamorphism is not related directly to the intrusion. It is not therefore an example of contact metamorphism and so presumably results from regional metamorphism.

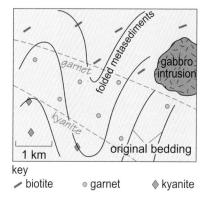

Figure 11.7   Geological sketch map, for use with the answer to Question 6.6.

Question 6.7

Point A is in a mafic igneous rock within the sillimanite zone, and so should contain amphibole and plagioclase feldspar and garnet (Table 6.1). Point B is in a calcic rock in the kyanite zone, and so should contain garnet, anorthite and amphibole. Point C is in a calcic rock in the garnet zone, and so should contain garnet, epidote and amphibole. Point D is in a calcic rock in the staurolite zone, and so should contain garnet, anorthite and amphibole.

Question 6.8

At point A, $T = 620$ °C and $P = 400$ MPa. This indicates a high geothermal gradient and, according to Figure 6.21, amphibolite facies conditions. (Near A, the metamorphic facies would increase rapidly with depth from greenschist near the surface to amphibolite and, ultimately, granulite facies.) At point B, $T = 290$ °C and $P = 1600$ MPa, corresponding to blueschist facies on Figure 6.21 and indicating a very low geothermal gradient.

Question 6.9

Water as either a liquid or a vapour is less ordered and thus has a higher entropy than the other phases, which are all (solid) minerals. The assemblage staurolite and quartz therefore has the lower entropy, so the assemblage garnet and sillimanite + H_2O, which has the higher entropy, must be stable at higher grades of metamorphism. Conventionally, an isograd is named after the index mineral on the side of *higher* metamorphic grade. Thus, this reaction would take place at the sillimanite isograd.

Question 6.10

The presence of staurolite indicates that the temperature was greater than 580 °C. As kyanite is present at what appears to be moderately high temperatures (>580 °C), the pressure must have been greater than 530 MPa (Figure 11.8). The fact that these pelitic rocks have not melted to form granites implies the temperature was less than 625 °C at these pressures. Thus, during metamorphism the temperature was between 580 °C and 625 °C at pressures greater than 530 MPa (see red shaded area on Figure 11.8). This assemblage is from the central Alps in Switzerland, and other evidence suggests that the pressures were 600–700 MPa.

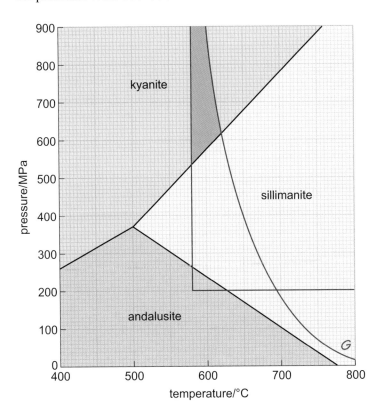

Figure 11.8 Phase diagram for answer to Question 6.10. Curve G is the water-saturated granite solidus.

Question 6.11

At 800 MPa and 200 °C, a rock belongs to the blueschist facies (Figure 6.21). 800 MPa corresponds to a depth of:

$$\frac{800 \text{ MPa}}{30 \text{ MPa km}^{-1}} = 26.7 \text{ km}$$

So (assuming a surface temperature of 0 °C) the geothermal gradient is:

$$\frac{200 \ °C}{26.7 \ km} = 7.5 \ °C \, km^{-1}.$$

This represents relatively high-pressure/low-temperature metamorphism of the kind typically found associated with subduction at convergent plate boundaries.

At 200 MPa and 550 °C, a rock plots in the hornfels facies near the transition into the amphibolite facies (Figure 6.21). This is relatively high temperature–low pressure metamorphism. 200 MPa corresponds to a depth of:

$$\frac{200 \ MPa}{30 \ MPa \, km^{-1}} = 6.7 \ km$$

So (assuming a surface temperature of 0 °C) the geothermal gradient is:

$$\frac{550 \ °C}{6.7 \ km} = 82 \ °C \, km^{-1}$$

which is much higher than the geothermal gradient predicted by models of crustal thickening (Figure 6.24, curve 2). It is most likely to be associated directly with magmatic activity in which hot magma has significantly increased the temperature at relatively shallow depths. At present, most magmas, other than those formed at divergent boundaries, are generated above subduction zones in island arcs, or above convergent continental plate boundaries, such as the Andes.

Question 7.1

No – generally not. The reason is that, although you might know which unit lies each side of the fault, you usually will not know what part of the unit lies at the ground surface. It could be the top or base of the bed that is exposed, or anywhere in the middle. Additionally, you may not know the exact thicknesses of the strata either side of the fault.

Question 7.2

(a) To answer this question, you need to measure the vertical distance from the base of the Hampen Formation to the top of the Birdlip Limestone Formation on the generalised vertical section, and apply the scale given next to the section. The distance on the vertical section is about 1.4 cm, which at a scale of 1 : 2500 equates to a throw of about 35 m for the fault (1.4 × 2500 = 3500 cm = 35 m).

(b) The distance on the vertical section is about 1.8 cm, which at a scale of 1 : 2500 equates to a throw of about 45 m for the fault at this point (1.8 × 2500 = 4500 cm = 45 m).

(c) The difference between the two answers (10 m) is substantial, and hence is unlikely to reflect the relatively small uncertainties in measuring from the vertical section. It is much more likely to indicate that the throw actually does vary along the length of the fault.

Question 7.3

(i) Repetition of strata across the faults. Moving from west to east across the faults, Proterozoic and Cambrian–Devonian strata are repeated initially, and then a succession of faults duplicates Carboniferous and Cretaceous strata. This duplication is typical of thrust faults.

(ii) Emplacement of older rocks (in the hanging wall) over younger rocks (in the footwall). Most (but not all) of the thrust faults in the southeastern portion of the map obey this general rule, though there are exceptions, especially where the age of strata in the hanging wall or footwall changes along the fault.

Question 7.4

Figure 11.9 shows the outcrop pattern after erosion for increased throw on the fault. Since a greater amount of layer 4 would have been removed by erosion, you would expect to see a wider sliver of layer 3 lying immediately south of the fault. Traversing from north to south across the two blocks, you would therefore expect to encounter beds 4, 5, the fault, a wider outcrop of 3, and then 4 and 5 repeated further southward on the southern, upthrown block.

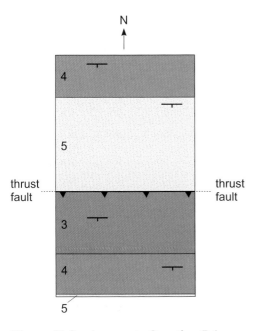

Figure 11.9 Answer to Question 7.4.

Question 7.5

The wavelength can be measured roughly from the anticlinal crest by the shore (lower right) to the plunging anticline hinge at the left end of the near cliff, a distance of about 55 m. The amplitude (half the height from anticline crest to syncline crest) cannot be measured easily on this photograph, but you could try tracing any single bed from an anticlinal crest to a synclinal crest. The amplitude must be at least 10 m (half the height of the near cliff), but is probably rather more than this.

Question 7.6

If dip symbols are given, then the strata forming an anticline will dip away from the fold axis, whereas those of a syncline will dip towards the fold axis. If valleys have been eroded into the folded strata, then another method of determining dip directions would be to examine the outcrop V-patterns. (Outcrop Vs of dipping strata in valleys tend to point in the direction of dip.) The anticline will contain the oldest strata in the centre (core) along the fold axis; a syncline will have the youngest strata exposed in its core.

Question 7.7

(a) This fold has the youngest strata in its core, so it is a syncline.

(b) The fold nose of this syncline points WSW, so the syncline must plunge in the *opposite* direction (i.e. towards ENE).

Question 7.8

Both are dip-slip faults in which the hanging wall has moved up relative to the footwall. However, thrusts have low fault-plane dips, which are less than 45° and typically less than about 20°, whereas reverse faults have steeper dips of over 45°.

Question 7.9

Normal faults have relatively steep fault planes, typically dipping at 60° or so, with mostly dip-slip movement in which the hanging wall is displaced down-dip. *Thrust faults* have relatively flat-lying fault planes, typically dipping 0–20°, with mostly dip-slip movement in which the hanging wall is displaced up-dip. *Strike-slip faults* have very steep fault planes with almost all horizontal movement and relatively little vertical movement. Displacement can be either to the right (dextral) or to the left (sinistral).

Question 7.10

Joints are discrete fracture planes through rock that separate blocks of undeformed rock. This is consistent with brittle deformation, but not with ductile deformation since that occurs without fracturing. Joints are therefore brittle structures.

Question 7.11

This is a dextral shear zone. The block of rock on which the lens cap sits has been displaced to the right relative to the other block, as shown by the clockwise swing of the foliation in the rock (on both sides of the zone).

Question 8.1

(a) In Figure 8.3a, the circle has a diameter of 30 mm. The long axis of the ellipse has a length of 45 mm. The extension is given by Equation 8.1:

$$e = \frac{l - l_0}{l_0} = \frac{45 - 30}{30} = \frac{+15}{30} = +0.5$$

From Equation 8.2, the *percentage* extension is given by:

$$\% \text{ extension} = e \times 100\% = +0.5 \times 100\% = +50\%$$

The plus sign shows that the line has lengthened during deformation.

(b) The short axis of the ellipse in Figure 8.3b is 20 mm long. Using the same method, % extension = −33%. The minus sign shows that the line has shortened during deformation.

Question 8.2

(a) The angle A′ D′ C′ in Figure 8.5b is 63°. Therefore, angle ψ, the shear, is 90 − 63° = 27°. From Equation 8.3:

$$\gamma = \tan \psi = \tan 27° = 0.5.$$

(b) As in Question 8.1, use Equations 8.1 and 8.2 and compare the post-deformation lengths of the ellipse axes in Figure 8.5b with the diameter of the circle in Figure 8.5a. The circle in that figure has a diameter of 30 mm. The long axis of the ellipse in Figure 8.5b has a length of 38 mm. The % extension is given by Equations 8.1 and 8.2:

$$e = \frac{l - l_0}{l_0} = \frac{38 - 30}{30} = \frac{+8}{30} = +0.27$$

From Equation 8.2, the percentage extension is given by:

$$\% \text{ extension} = e \times 100\% = +0.27 \times 100\% = +27\%.$$

(c) The short axis of the ellipse in Figure 8.5b is 23 mm long. Using the same method as in part (b) above, % extension = −23%.

Question 8.3

The long axis of the ellipse is 30 mm and the short axis is 15 mm. The axial ratio of this ellipse is the ratio of the long axis to the short axis, i.e. 30 : 15 or 2 : 1.

Question 8.4

The time taken for an extension of 1, i.e. $\frac{2d - d}{d} = \frac{d}{d}$, at a strain rate of 10^{-10} s^{-1}, would be $\frac{1}{10^{-10} \text{ s}^{-1}}$ which is 10^{10} s. Using a figure of 3×10^{13} seconds in 1 Ma, the Alps would have doubled in thickness in about 3×10^{-4} Ma, or around 300 years. So the strain rate observed at Etna really is staggeringly fast, and such strain rates are probably only sustained for geologically short periods.

Question 8.5

The angular shear, ψ, is the change in angle between two lines originally at right angles. The bilateral symmetry of the trilobite can be used to give lines originally at right angles. From Figure 8.6, the angle between the central axis of the fossil and the horizontal body segments, which was originally a right angle, is now about 45°. Therefore, angle ψ, the shear, is 90° − 45° = 45°, so the shear strain $\gamma = \tan \psi = \tan 45° = 1$. You cannot measure the extension of, say, the length of the trilobite, because you do not know how long the sheared specimen was before deformation (even though you know how long its friend is).

Question 8.6

Below is the completed description:

'These siltstones initially deformed in a ductile manner, by <u>folding</u>, before further strain resulted in brittle <u>fracturing</u>. The hanging wall of the thrust has been <u>translated</u> from right to left. The steep limbs of the folded beds have been <u>rotated</u> from their original horizontal orientation. The fact that once-planar beds are now curved and folded shows that they have been <u>distorted</u>.'

Question 9.1

Normal faults, thrust faults and joints are all brittle structures and are likely to form in the upper parts of the crust. Ductile shear zones are more likely to form at depth.

Question 9.2

(a) A homogeneous body of rock could simply change its shape uniformly, so that an original cube became shorter in one direction and thicker at right angles to that direction. Rotation of grains and pressure dissolution would produce a slaty cleavage.

(b) In a heterogeneous body of rock, less competent units may shorten as in (a), but more competent units would form buckle folds, with or without a cleavage.

Question 9.3

A parallel fold is one in which the thickness of the folded layer remains constant right around the fold. It implies the layer was competent, relative to the layers around it. Not many layers can form parallel folds before a space problem arises in the fold core. A similar fold is one in which all folded surfaces have the same curvature. They generally form in incompetent, mobile material and require significant volume changes along individual layers. Similar folds can be stacked together indefinitely.

Question 10.1

The % extension is −30%. The present-day, straight-line distance, l, between the ends of the middle bed of the Cretaceous succession, from the WNW end of the section near the Ratz anticline to ESE of the Chartreuse Hills, is approximately 21 km (adjusting for the horizontal scale on the cross-section). Measuring the actual length of that same stratum, around all the folds and across the faults, even where it has now been eroded away, gives a pre-deformation length, l_0, of about 30 km. That stratum was initially 30 km long – now it is only 21 km long. The extension, e, and percentage extension can now be calculated from Equations 8.1 and 8.2.

$$e = \frac{l - l_0}{l_0} = \frac{21 \text{ km} - 30 \text{ km}}{30 \text{ km}} = \frac{-9 \text{ km}}{30 \text{ km}} = -0.3$$

$$\% \text{ extension} = e \times 100\%$$
$$= -0.3 \times 100\%$$
$$= -30\%$$

Acknowledgements

In addition to those mentioned in the Course Team list, the authors would like to thank Helen Craggs for comments on the proofs.

Grateful acknowledgement is made to the following sources for permission to reproduce material in this book.

Figures

Cover: Andy Sutton.

Figures 1.1a, 3.1, 3.2a, 3.4, 3.12a, 3.24 and 4.12a: David Rothery; Figure 1.1b: Professor Chris Hawkesworth; Figures 2.3a, 4.5, 4.16a, 5.13c and 10.12: Reproduced with the permission of the British Geological Survey © NERC. All rights Reserved; Figure 2.3b: Adapted from Ekman, M. (1996) 'A consistent map of the postglacial uplift of Fennoscandia', *Terra Nova*, vol. 8, 1996, 158–165. Blackwell Science Limited; Figures 2.12, 3.3, 3.12c, 3.13a, 3.29, 7.11 and 7.27: United States Geological Survey ; Figures 3.2b and c: Adapted from Flynn, L.P., Mouginis-Mark, P.J., and Horton, K.A. (1994) 'Distribution of thermal areas on an active lava flow field: Landsat observations of Kilauea, Hawaii, July 1991', *Bulletin of Volcanology*, 56(4), 284–296. Springer Publishing Company; Figure 3.2d: Kathy Cashman, Oregon State University/USGS; Figures 3.5, 3.7b, 3.13c, 3.16, 3.20 and 3.33b: Stephen Blake: Figures 3.7a and 5.13a: Peter Francis; Figure 3.7c: Landsat image courtesy of the U.S. Geological Survey; topography derived from modified SRTM data: Jarvis, A., H.I. Reuter, A. Nelson, E. Guevara, 2008, Hole-filled seamless SRTM data V4, International Centre for Tropical Agriculture (CIAT); Figures 3.8 and 3.23: Kilburn, C.R.J. (2000) in Sigurdsson, H. et al. *Encyclopedia of Volcanoes*. Academic Press; Figures 3.9, 6.2, 6.4c, 6.7, 6.20a, 10.5 and 10.9: Nigel Harris; Figure 3.12b: Marco Fulle-www.stromboli.net; Figure 3.13b: Texas Beyond History, Texas Archeological Research Laboratory, University of Texas at Austin; Figure 3.13d, 6.19 and 6.20b: Andy Tindle; Figure 3.14: Richard P. Hoblitt/United States Geological Survey; Figure 3.17: Sigurdsson, H. et al. (1985) 'The eruption of Vesuvius in AD79', *National Geographic Research*, 1(3), 332–387. National Geographic; Figures 3.18 and 4.1a: NASA; Figure 3.19: Adapted from Sparks, R.S.J. et al. (1997) *Volcanic Plumes*. John Wiley and Sons Limited; Figure 3.26: Professor Tim Druitt, Université Blaise Pascal; Figure 3.27a: Smithsonian Museum of Natural History, http://www.si.edu/; Figure 3.27b: N Banks/USGS; Figure 3.28: Voight, B. and Cornelius, R.R. (1991) 'Prospects for eruption prediction in near real-time', *Nature*, vol. 350, issue 6320, 695–698. Nature Publishing Group; Figure 3.29, 7.11 and 7.27: United States Geological Survey; Figure 3.31: Ui, T. et al. (2000) in Sigurdsson, H. et al., *Encyclopedia of Volcanoes*, Academic Press; Figure 3.32 Adapted from Williams, H. (1941) 'Calderas and their origins'. *University of California Publications, Bulletin of the Department of Geological Sciences*, 25, 239–346. University of California; Figure 3.33a: Adapted from Aldiss, D.T. and Ghazali, S.A. (1984) 'The regional geology and evolution of the Toba volcano-tectonic depression, Indonesia', *Journal of the Geological Society of London*. 141, 487–500. The Geological Society, London; Fig 4.1b Haflidason, H. et al. (2000) 'The tephrochronology of Iceland and the North

Index

Entries in **bold** represent glossary terms. Page numbers in *italics* refer to figures and tables..